GO HERE INSTEAD

THE
ALTERNATIVE
TRAVEL LIST

GO HERE INSTEAD

The mighty Annapurna mountain range in the Himalayas, Nepal

CONTENTS

Clockwise from top right
Cruising through pack ice in North Greenland; kob in the savannah, Western Uganda; dancing during the Montevideo Carnival, Uruguay

INTRODUCTION

We see that bucket list of yours (dog-eared with time and filled with far-away places) and we raise you: one alternative travel list.

New York City, Angkor Wat, the Great Barrier Reef: everyone wants to visit them, and with good reason. But it's time to dream of somewhere different. Our world is packed with incredible places, places we often miss if we stick to the tourist trail. We want to shine the spotlight on them for a change.

Why stray from the beaten path? Aside from the joy of seeking out somewhere new, we have a couple of reasons. Firstly, fewer people. Fame comes at a cost and the most iconic sights often come with equally famous crowds (Venice, we're looking at you). Secondly, choosing to swerve the big hitters might mean choosing a more local and sustainable experience – so you'd be doing yourself, and our planet, a favour.

To get you going, we've made *Go Here Instead: The Alternative Travel List*. Inside this book, we've rounded up an array of incredible alternatives to the world's most visited places – because for every iconic sight there's a lesser-known rival. We're thinking the northern powerhouse giving London a run for its money, the island blooms just as beautiful as Japan's cherry blossoms and the fun-filled festival taking the shine off Coachella.

And that's just a taster. Flick through these pages and you'll discover everything from great journeys few embark on to dazzling natural wonders overlooked by the masses. You might have heard of a few of the sights inside this book – when is anything completely off the beaten track? – but we bet you'll still find plenty of hidden gems to inspire your future travels. So, don't stick to what you know, take the road less travelled instead.
The world is waiting.

ANCIENT AND HISTORICAL SIGHTS

The monumental Stonehenge has been perplexing visitors for centuries but it's not the only Neolithic site with a legendary past. Discover the equally head-scratching (and much bigger) Avebury, replete with its own myths and mysteries.

the alternative to *Stonehenge, UK*

AVEBURY

UK

Crowning the skyline on the open expanse of Salisbury Plain are the jaw-dropping prehistoric standing stones of Stonehenge. Coach after coach disgorges load after load of snap-happy tourists, who then complete their circuit of the stone circle – from a distance (Stonehenge is so popular that the site has to be roped off for fear of damage).

To really get hands-on with ancient stones, and to see the world from a 5,000-year-old perspective, head to Avebury, located 20 miles (33 km) to the north of Stonehenge. This mystic circle of stones, or henge, was erected around the same time, but it continues to fly under the radar of the coach-trip hordes. Avebury's stones are not as huge as those at Stonehenge, but the site is four times the size, making it the world's largest henge. Within a ditch that measures nearly 2 km (a mile) in circumference, the remaining giant gnarled thumbs of sandstone – astonishing in scale – are complemented by concrete blocks showing how the circle and its outer and inner rings once stood.

Over the years, the village of Avebury itself has grown up around (and within) the henge. The stones occupy what appears to be a village green, with roads cutting through the centre and several houses (plus a pub) dotted alongside. It gives the henge the appearance, at least from the air, of an ancient pie chart. You can visit at any (reasonable) time, a laid-back approach that both befits the surroundings and is a far cry from the scheduled shuttlling at Stonehenge.

Avebury's giant sarsen stones, dating to between 2850 BC and 2200 BC

Above *The ancient stones surrounded by sheep against the rising sun* ***Right*** *The Neolithic stone circle and Avebury village, which lies mainly within the monument*

Like the resident sheep, you can wander freely among the stones at Avebury, touching them and seeing if some of their mystery will rub off on you. It's powerful stuff, a bit too powerful for some, perhaps – don't be surprised if you see people hugging them. As with Stonehenge, nobody really knows the purpose of these strange circles but, like most great constructions of early civilizations, they were likely part of a culture that combined religion with the heavens and afterlife.

Since the Middle Ages, the stones have been wreathed in superstition. At the centre of the northern circle lie the remains of the Cove, also known as the Devil's Brand-Irons, and other stones here have names like the Devil's Quoits and the Devil's Chair. It's probably no coincidence that a church was built right next to the henge, and it's thought that there are more stones beneath Avebury, buried by villagers who were fearful of the circles' pagan powers. Later, in the 17th and 18th centuries, locals quarried the henge for building materials, breaking up the stones to use them in the walls of houses. Legend has it that these homes were haunted. The village's entire history is mixed up with the stones, in fact, and can be explored in the local museum set up by amateur archaeologist Alexander Keiller, who removed buildings from the site and re-erected many stones in the 1930s.

Before you leave Avebury, head to a couple of other linked Neolithic sites nearby: about 2 km (a mile) southeast at the end of West Kennet Avenue, there's a path flanked with a stone circle, called the Sanctuary. A little less than a mile to the west of here is Silbury Hill, a conical mound made at about the same time as the henges, the purpose of which has not

been fully ascertained either. Local folklore says that the earth that makes up the mound was dumped here by the Devil.

Whether you believe that or not, whether you think that Avebury was built as a theatre for sacrificial ceremonies or as an astronomical clock, or is just an interesting circle of old stones, you'll have plenty of time and space to explore.

Still going to Stonehenge?

Book for English Heritage's memorable one-hour Stone Circle Experience. Unlike the standard entry, this lets you venture inside the sarsens and visit at quieter times of the day, before or after the normal opening hours.

Getting There *It's a one-hour train journey from London Paddington to Swindon, from where bus 49 takes half an hour to reach Avebury.*

www.english-heritage.org.uk/visit/places/avebury

RING OF BRODGAR
UK

In the Heart of Neolithic Orkney heritage site in Scotland, the circle of tall standing stones of Brodgar dates to 2,000 BC. This beautiful place also has burial mounds and earthworks. The Stones o' Stenness, an earlier henge, is less than a mile away.

ALES STENAR
Sweden

This monument, dating to AD 500–1000, sits atop a high moraine on the edge of the Baltic Sea in southern Sweden. It is the most significant of Sweden's "ship settings" – rings of standing stones in the shape of a boat.

WASSU STONE CIRCLES
The Gambia

It's believed that kings and chiefs were buried among the 11 stone circles of Wassu about 1,200 years ago. These are the best examples of hundreds of similar such monuments found in the Senegambia region of The Gambia.

Often overlooked in favour of Giza's iconic monuments, the pyramids of Saqqara and Dahshur are even older. You won't find endless processions of tour buses or touts here, just desert and awe-inspiring ruins.

the alternative to *the Pyramids of Giza, Egypt*

PYRAMIDS OF SAQQARA AND DAHSHUR

Egypt

The entrance to the complex of the monumental Step Pyramid of Zoser at Saqqara, the oldest known complete stone structure in the world, built around 2650 BC

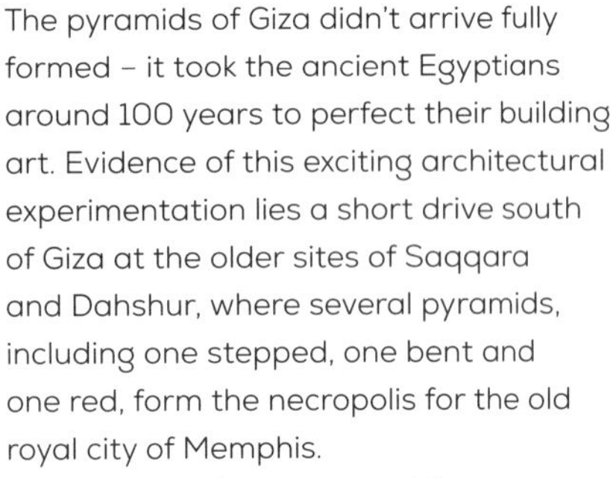

The pyramids of Giza didn't arrive fully formed – it took the ancient Egyptians around 100 years to perfect their building art. Evidence of this exciting architectural experimentation lies a short drive south of Giza at the older sites of Saqqara and Dahshur, where several pyramids, including one stepped, one bent and one red, form the necropolis for the old royal city of Memphis.

The idea of using pyramids as mausoleums was born during the Third Dynasty of ancient Egypt, and the funerary complex at Saqqara contains five original structures. Dominating the site is the Step Pyramid of Zoser – the first Egyptian pyramid and the tomb of Pharaoh Zoser. The royal sarcophagus was sunk into the bedrock of the desert, then surrounded by a granite platform that was gradually expanded into a four- and then six-step pyramid according to very precise calculations. The pharaoh's engineer, Imhotep, oversaw the design and construction of this pyramid so well that he was later raised into the pantheon of the gods, an honour only bestowed on a handful of people in ancient Egypt. Although smaller pyramids surround Zoser's resting place, the tomb of Imhotep (the world's first-named architect) has never been found.

The pharaohs of the Fourth and Twelfth Dynasties chose Dahshur, to the south of Saqqara, as the site for their cemetery. Leading the way was the ruler, Snefru, who ordered the building of Dahshur's two most magnificent pyramids, the Red Pyramid and the Bent Pyramid, and kick-started the golden age of Egyptian pyramids in the process. The Red Pyramid, named after the colour of its limestone blocks, is thought to be the first true pyramid, with sloping slides

Top *The striking Bent Pyramid at Dahshur, whose upper half was built at a lower angle than the bottom* ***Above*** *A relief inside a tomb at Saqqara*

rather than steps and descending passages through the blockwork to tomb chambers. The construction techniques took time to hone, though, since the nearby Bent Pyramid displays a change in angle halfway up one side. It also, strangely, has two separate entrances.

Snefru's son Cheops, inspired by his father's architectural efforts at Dahshur, went on to build the Great Pyramid of Giza, which was finished around 2650 BC. While the Giza pyramids may have graced postcards for centuries, let's not forget that those that paved their way at Saqqara and Dahshur are equally awe-inspiring structures.

Still want to see the Pyramids of Giza?
Visit in winter to avoid the summer heat. Don't miss the Solar Boat Museum, with a reconstruction of Cheops's wooden boat.

Getting There *Saqqara and Dahshur are respectively 25 km (16 miles) and 35 km (22 miles) south of Cairo, making an easy day trip from the city.*

www.egypt.travel

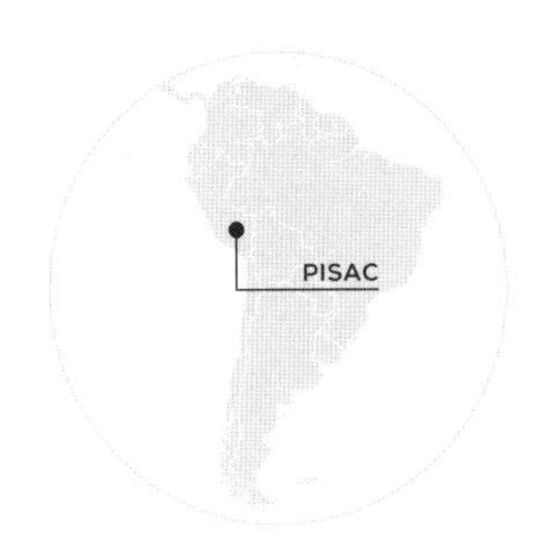

Machu Picchu isn't all the Inca Empire left behind. High above the Sacred Valley, the beguiling ruins of Pisac take you back half a millennium to when the Inca Empire was at its zenith.

the alternative to *Machu Picchu, Peru*

PISAC

Peru

The dramatic ruins of Machu Picchu may be Peru's ultimate drawcard, but they overshadow a whole host of other ancient Inca treasures – Pisac being one of them. This tranquil market town lies around 45 km (28 miles) southeast of Machu Picchu, and sits at the foot of a hill covered by another set of magnificent Inca ruins. Despite being only a short drive from Cusco, the former Inca capital, it remains under the radar.

Dating back more than 500 years to a time when the Inca ruled the largest empire in the Americas, the hilltop Pisac citadel is still remarkably intact. Get a sense of what life was like here in its heyday as you wander through the archaeological park, past preserved remains of temples, residential quarters, tombs, and ceremonial baths fed by an ingenious system of channels and aqueducts. The steep slopes below the site are lined with terraced fields, an ancient agricultural technique perfected by the Inca and still in use throughout the central Andes. From the precipitous summit, the views across the Sacred Valley are nothing short of amazing. Machu Picchu, you've got competition.

Still set on seeing Machu Picchu?
Although you'll have plenty of company (Machu Picchu is one of South America's most popular attractions), the glorious Inca ruins still take the breath away. There's now a cap on the number of visitors, so entry tickets should be booked about six months in advance.

Getting There *The nearest airport to Pisac is in Cusco, 35 km (22 miles) southwest, where you can pick up a bus or taxi.*

www.peru.travel

Pisac's evocative Inca terraces stretching over the verdant hillside, with the Sacred Valley beyond

The Colosseum might be the most famous Roman arena, but Nîmes' amphitheatre vividly evokes the grandeur of the Roman Empire at the pinnacle of its power, without a pandemonium of tourists.

the alternative to *the Colosseum, Italy*

LES ARÈNES DE NÎMES

France

Arriving at the 2,000-year-old Arènes de Nîmes, it's easy to imagine toga-clad spectators cheering while gladiators fought each other or wild animals. And hasn't that always been the point of amphitheatres: to entertain the masses? Les Arènes de Nîmes continues to enthrall, with (albeit tamer) concerts and festivals, recalling the jovial spirit of Roman times.

Performances aside, Les Arènes de Nîmes' fascinating history is reason enough to visit – think of the Visigoths who transformed the site into a fortress, or the peasant housing that filled it during the Middle Ages. Its story continues to evolve, too, with excavations uncovering subterranean surprises. The Colosseum in Rome may be the largest in the ancient world, but you'll leave Les Arènes de Nîmes with just as grand stories to tell.

Still set on the Colosseum?

The architecture is truly impressive, but it's the stories of the emperors and gladiators who graced this amphitheatre that you'll want to hear about, so book a guided tour.

The intricate exterior of Les Arènes de Nîmes, one of the best preserved of all Roman amphitheatres

Getting There *TGV trains from Paris (Gare de Lyon) take about three hours to reach Nîmes Centre train station, southeast of Les Arènes de Nîmes.*

www.arenes-nimes.com

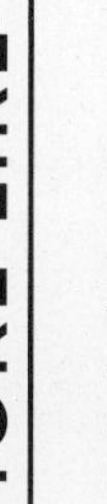

ARENA DE VERONA
Italy
Encircled by two decks of stone arches, this elegant Roman amphitheatre is now a renowned venue for excellent opera performances.

AMPHITHÉÂTRE D'EL-JEM
Tunisia
This remarkably well-preserved 3rd-century amphitheatre could once seat 30,000 spectators. Its tribunes were held aloft by three tiers of stone barrel arches.

Easter Island's moai *may be the world's most iconic stone heads, but they lie on one of the remotest places on earth. Enter the colossal statues of Turkey's Nemrut Dağı – much more accessible, and just as awe-provoking.*

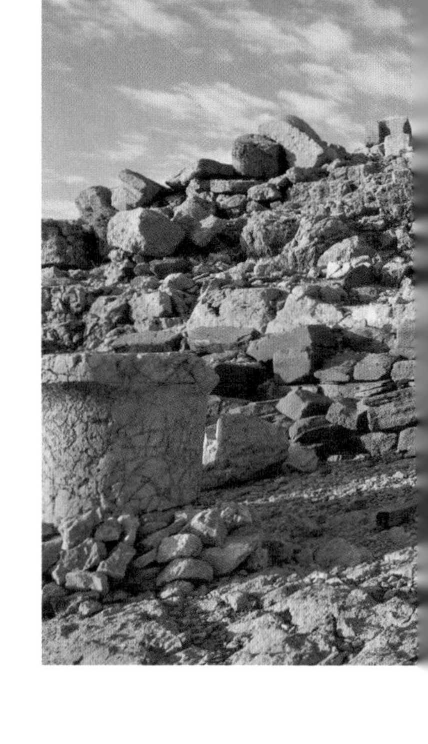

***the alternative to** the* moai *of Easter Island, Chile*

STONE HEADS OF NEMRUT DAĞI

Turkey

It's labelled "Turkey's Easter Island", but Nemrut Dağı is a world unto its own. A towering peak in the Eastern Taurus mountain range in southeastern Turkey, the 2,134-m (7,001-ft) Nemrut Dağı is home to a set of giant statues that rival the monolithic *moai* of far-flung Easter Island. Although relatively remote itself, Nemrut Dağı is easily visited on a day trip from many villages, towns and cities. Easter Island, by contrast, is more than 3,500 km (2,175 miles) west of mainland Chile, and getting there involves a pricey 4.5-hour flight from the capital Santiago. Do the maths, and Nemrut Dağı comes out on top.

Crowning Nemrut Dağı's summit are the ruins of an ancient tomb and temple built for King Antiochos I of Commagene (69–34 BCE), who once ruled over much of the surrounding region. The highlights up here, of course, are the looming statues on terraces around Antiochos' burial mound, a 60-m- (197-ft-) high pile of rocks. Over time, earthquakes have decapitated these stone giants, leaving their heads, around 2 m (6.5 ft) high, resting on the ground at their feet. It makes for a strangely moving, melancholy sight.

Antiochos would often boast that he was descended from both Alexander the Great and Darius the Great of Persia, and the design of the statues reflects the union of ancient Greek and Persian cultures that his kingdom embodied. By carving Greek-style facial

Colossal stone head on the west side of the summit of Nemrut Dağı

features with Persian hairstryles and clothing, gods and heroes from both cultures are merged: Greek god Zeus with Iranian god Oromasdes, Heracles and Ares with Indo-Iranian deity Artagnes. There are broken statues of a lion and an eagle, too, who acted as guardians of the monument, and another of Antiochos.

Being here, the mind boggles at the effort it must have taken to transport the stones up the mountain, let alone the imagination required to carve them into such beguiling forms. A UNESCO World Heritage Site since 1987, Nemrut Dağı is a glorious feat of iconography.

Stone head looking out from the East Terrace of Antiochos I's burial mound on Nemrut Daği

Still set on seeing Easter Island?
Although flights are expensive, Easter Island is a magical place to visit. There are countless *moai* here so, as well as popular spots like Rano Raraku, head to the lesser-visited Ahu Akivi and seek out the lone statue on Pia Taro road.

Getting There *Guided tours to Nemrut Daği can be organized in many towns in the surrounding area, including Malatya and Kahta.*

www.goturkiye.com

SAN AGUSTÍN ARCHAEOLOGICAL PARK
Colombia
This UNESCO World Heritage Site in southwestern Colombia is home to the largest collection of ancient megalithic sculptures and religious monuments in South America, including innumerable stone statues.

TIWANAKU
Bolivia
Near Lake Titicaca, the ruined capital of the pre-Inca Tiwanaku culture has a subterranean temple covered with scores of limestone heads thought to represent the deities of vanquished foes. Each carved face is entirely unique.

BADA VALLEY
Indonesia
On the island of Sulawesi, this valley is studded with around 400 ancient megalithic statues, which were made by an unknown culture perhaps as long as 5,000 years ago. The stone statues have very minimalistic designs.

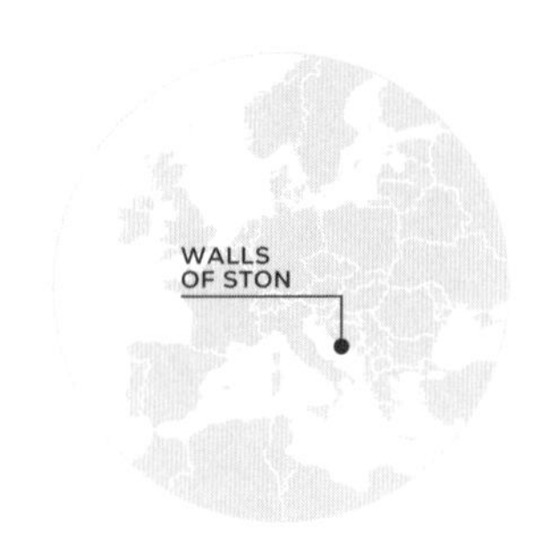

China's world-famous wall may be unmatched for length, but most of it only dates from the 15th century. Croatia's Walls of Ston, meanwhile, were begun in 1358 – and you can still walk most of the original construction.

The rising sun hitting the long, zig-zagging Walls of Ston and the surrounding verdant hills

the alternative to *the Great Wall of China, China*

WALLS OF STON

Croatia

Tying up the northeastern edge of Croatia's Pelješac peninsula like an unbreakable bow, the stunning Walls of Ston – known colloquially as "Europe's Great Wall" – are the second-longest defensive barrier in the world. The first-longest (yes, China's) may hog the limelight, but the Walls of Ston have just as many headline-grabbing hallmarks, not least the fact that a looped walk only takes three hours – compare that to countless months spent trekking the Great Wall of China – without breaking a sweat.

Stretching strategically from coast to coast, Croatia's defensive ramparts were built to protect the area's valuable salt pans from invaders and pillagers.

Walking along the cobbled defence wall past lush greenery towards one of the remaining towers

Naturally, the job was entrusted to some of Europe's finest Renaissance minds, including the famed Florentine architect Michelozzo. There were a total of 40 towers and five fortresses situated all along the 7-km (4.5-mile) span, and the wall was top-and-tailed by two fortified towns, Ston and Mali Ston, built to house the border guards and salt-pan workers.

Today, almost all of the original wall survives, along with around half the towers. Ston marks the starting point for a unique wall-top walk, where the fortification gracefully arcs up the hillside, beckoning you to clamber aboard and follow its path. The view expands with every step, unveiling a spectacular landscape of lush green forests overlooking the quaint orange roofs and shimmering salt pans of Ston. When you breach the brow of the hill, the chevron-shaped Kiruna Fortress

silently guards the northern section of wall. Venture down to Mali Ston to sample its fresh oysters, or stop to soak up the views of the ancient structure surrounding you. After all, it's not every day you find yourself standing on Europe's Great Wall.

Still going to the Great Wall of China?
The world's most famous fortification attracts around 10 million tourists each year, and most of them head for the best-preserved sections near Beijing. For a more serene stroll, venture a little further to the seaside Shanhai Pass – the eastern edge of the Ming dynasty Great Wall.

Getting There *Buses run from the city of Dubrovnik to Ston three times a day and take 75 minutes.*

www.ston.hr

MASADA

Ancient fortified palaces built by uneasy monarchs crown both Sigiriya and Masada, but only the latter was the site of a valiant but doomed last stand against the Roman Empire.

the alternative to *Sigiriya, Sri Lanka*

MASADA

Israel

Sigiriya might have the looks, sitting atop a giant gneiss rock surrounded by lush forest, but Masada has the stories (its location above the Dead Sea isn't too shabby, either).

Fortified in the 1st or 2nd century BC, Masada was enlarged into a palace complex by Herod the Great, King of Judea. The Romans took Masada when Herod died, but in AD 66 it was captured by Jewish rebels during the First Revolt. Judea's last Jewish rebel stronghold, Masada was under Roman seige for over two years until the walls were breached in AD 73. As defeat loomed, the rebels chose mass suicide over submission to Rome. Discover traces of this remarkable event here today, including excavated remains of a synagogue and Roman military camps.

Still want to see Sigiriya?
It's not all about reaching the top: spend some time away from the crowds ogling the reservoirs and gardens at the base.

Getting There *Ben-Gurion International Airport, near Tel Aviv, is about two hours by car from Masada. A bus also runs from Jerusalem.*

www.parks.org.il

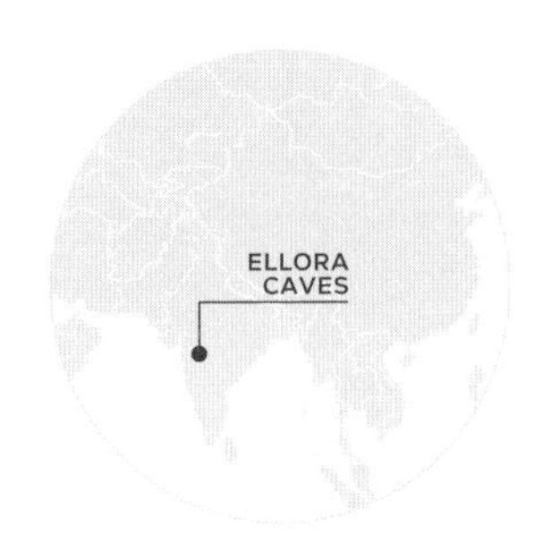

Scooped out of sandstone cliffs, the ancient city of Petra is undeniably a jaw-dropping sight. But for equally amazing rock-hewn buildings, opt for the Ellora Caves in India. These stunningly embellished temples will leave you speechless.

the alternative to *Petra, Jordan*

ELLORA CAVES

India

Carved out of the basalt cliffs of the Charanandri hills in Maharashtra state, western India, is one of the biggest complexes of rock-cut temples anywhere in the world, the Ellora Caves. These magnificent structures, chiselled between the 6th and 9th centuries, abound with audaciously elaborate decorations and immaculately carved sculptures. Yet they receive only about 21,000 visitors per year compared to the million or more who converge on Petra in Jordan.

Representing three of the world's most ancient religions – Hinduism, Buddhism and Jainism – the Ellora Caves served as temples for prayers, monasteries for monks and refuges for pilgrims to stay. These places of worship co-existed peacefully with each other for centuries. The Hindus wrapped their temples with figures of gods and goddesses from their pantheon. The Buddhists carved out *virāhas* (monasteries) and *chaityagrihas* (prayer halls) and decorated them with sculptures of serene Buddhas and Bodhisattvas. The Jains, meanwhile, sculpted sanctuaries and filled them with paintings depicting flying celestial beings. Petra's temples and tombs have monumental entrances but it's the intricacy of the decorations on the ceilings and the walls, the exquisitely crafted statuary and beautiful mural art that give the caves at Ellora the edge.

Of the 100 temple-caves, 34 are open to the public. The most stupendous is the magnificent Kailash Temple. Commissioned by the Rashtrakuta king Krishna I in the 8th century, this mammoth complex, complete with courtyards, passages, rooms and an 83-m- (272-ft-) high tower, was carved out of one huge rocky cliff face. Sculptors chiselled through

The Kailash Temple, the largest of the rock-cut Hindu temples at Ellora, formed from a single block of stone

A sadhu (Hindu holy man) in front of the Kailash Temple, dedicated to Lord Shiva

85,000 m³ (about 3 million ft³) of rock for over a hundred years, beginning at the top of the cliff and working their way down - no small accomplishment. The resulting marvel, adorned with huge sculptural panels, was meant to depict Mount Kailasa, the sacred abode of Lord Shiva.

After exploring the cave's interior, walk along the path to the south of the complex. This will take you to the top perimeter. From here, you'll have a bird's-eye view of all of the Ellora cave-temples – a thrilling sight.

Still going to Petra?

For the best photos, visit the monastery and royal tombs late in the afternoon when they're bathed in a pre-sunset glow. Beware of hawkers; should you wish to ward them off, you might want to learn the Arabic for "no, thank you": *la shukran.*

Getting There *Flights from India's Mumbai International Airport take about an hour to Aurangabad. From here buses run every half hour to the Ellora Caves.*

www.asi.nic.in/ellora-caves

LALIBELA
Ethiopia

This extraordinary complex of 11 churches, hand-sculpted out of red volcanic rock around the 12th century, is Ethiopia's number one tourist attraction. Pilgrims flock to Lalibela from miles around as it's still in use for worship to this day.

PHRAYA NAKHON CAVE
Thailand

A splendid gold-and-green royal pavillion is hidden inside this cave, 430 m (1,410 ft) up a mountain in Thailand's Khao Sam Roi Yot National Park. There are superb mountain and forest views from the cave.

DAMBULLA CAVE TEMPLE
Sri Lanka

Spread across five caves in the very centre of Sri Lanka, this 1st-century temple complex contains over 150 statues of Buddha and more than 1,500 paintings. From the caves there are excellent views over the surrounding countryside.

With its coffered ceilings and exquisitely carved archways, the Palacio de la Aljafería in Zaragoza is one of Spain's most beautiful Moorish buildings. And, unlike its iconic southern counterpart the Alhambra, it is still very much in use today.

the alternative to the Alhambra, Spain

ALJAFERÍA

Spain

Cresting a rocky hill in the old Albaicín neighbourhood of Granada, and backdropped by the snow-capped peaks of the Sierra Nevada, the Alhambra is Spain's most lavish Islamic building, and one of the most popular tourist attractions on the Iberian Peninsula. But did you know that at the opposite end of Spain, in the mostly mountainous region of Aragón, there's another extravagant palace? Built in the second half of the 11th century, a good 150 years before the Alhambra, the Palacio de la Aljafería in Zaragoza is an equally stunning Moorish building. Over the centuries, it's served as an Islamic summer palace, a home to Christian kings and the seat of the Spanish Inquisition. It's now the rather grandiose home of the Cortes de Aragón, the regional parliament.

Guided tours work their way through the Aljafería's eras. The starting point is the impressively manicured Patio de Santa Isabel, the gardened central courtyard of the original Islamic Palace, built by al Muqtadir, the ruler of Saraqusta (Zaragoza), in around 1065. Surrounded by flamboyantly carved arches, it's an incredibly tranquil spot (despite being in the centre of the city), with songbirds twittering in the lofty trees.

Santa Isabel has a touch of the Alhambra's Patio de los Leones about it, and there are more similarities ahead. Ducking through the arches on the courtyard's northern side you'll enter the Salón Dorado, or Golden Hall, one of the most sumptuously decorated areas of the palace. The rooms here, which were once the royal families' private bedrooms, have richly worked archways and coffered ceilings that, like the ceiling of the Alhambra's Throne Room, are decorated with the stars of the Muslim cosmos.

Moving upstairs, the tour passes through the medieval Christian Palace, which served as the 14th-century royal residence of the Aragonese kings, before arriving at the Catholic Monarchs' Palace. Constructed in the 15th century, with yet more delicious decoration, this palace boasts the Throne Room as its highlight.

Clockwise from above *The imposing exterior of the Aljafería, built in the 11th century; close-up of a carved archway; Patio de Santa Isabel, the central courtyard of the original Islamic palace*

Here you'll find another coffered ceiling, gleaming gold and polychrome. Its exquisitely carved squares and octagons, from which golden pine cones hang, are complemented by the geometric tiled floor. Indeed, architectural touches like these have ensured that the Aljafería is listed as a UNESCO World Heritage Site.

It's a thrilling journey through nearly a thousand years of history. And, back on the ground floor, in the area of the palace that's been refitted to house the regional parliament, history continues to be made.

Still set on visiting the Alhambra? Do the opposite of what most people do and start your visit at the Nasrid Palaces before moving on to the Generalife and the Alcazaba.

Getting There *Zaragoza International Airport is 16 km (10 miles) west of the city. Buses from the airport take 20–30 minutes to reach the city centre.*

www.cortesaragon.es

MORE LIKE THIS

ROYAL MONASTERY OF SANTA MARÍA DE GUADALUPE
Spain
This glorious complex in southwestern Spain includes a 15th-century Moorish church and cloister.

CASA DE PILATOS
Spain
Set around a central patio and delightful gardens, this 16th-century palace in Seville mixes Moorish, Renaissance and Baroque styles.

The Acropolis of Athens may be iconic, but it sits bang in the centre of a teeming metropolis. If you prefer your ancient temples with sea views and spectacular surroundings, then you'll be far better off in Sicily.

***the alternatives to** the Acropolis of Athens, Greece*

AGRIGENTO AND SELINUNTE

Italy

There's no getting away from it: the Acropolis of Athens is the most famous ancient site in Europe. But fame comes at a cost, and the rocky platform on which it stands is polished by the feet of four-and-a-half million visitors a year. And although Athens is no longer the most polluted city in Europe, it is still a very busy one, and the Acropolis stands right in the middle of it.

Driving through Sicily to the temples of Selinunte and Agrigento couldn't be more different. The landscapes here have changed little in two millennia – rolling hills blanketed with wheat fields, olive groves and vineyards. In fact, it was this fertility that attracted the Greeks to Sicily in the first place. Greece's soil had been rendered nigh infertile by deforestation (building ships and smelting metal used enormous amounts of wood), so the Greeks turned their sights elsewhere.

Founded in 581 BC, Agrigento (Akragas in Ancient Greek) became one of the wealthiest cities in Magna Graecia, the ancient Greek Empire. It was synonymous with excess, a place where enemies of the ruling tyrant Phalaris were roasted to death inside a bronze bull, and where the wealthy furnished their homes with ivory, wore abundant silver and gold, and even made elaborate tombs for their pets. By the 4th century BC, Sicily was the gastronomic capital of the Greek empire, notorious for decadent feasts featuring baby-bird pie and casseroled eel.

The well-preseved Tempio della Concordia in the Valley of the Temples at Agrigento

Above *Standing inside the Tempio di Hera (Temple E) at Selinunte* ***Right*** *Tempio di Hera on the Acropolis of Selinunte, surrounded by olive trees*

The gold, ivory and eel have long disappeared, but the Valley of the Temples (as Akragas is known) remains beautiful, set high on a ridge above the shimmering Mediterranean.

The temples are equally spectacular. The Tempio della Concordia is one of the best-preserved Greek temples in the world (and arguably on a par with Athens's Parthenon), though you may find the slightly less perfect Tempio di Giunone more romantic. There are more temples in the less-visited western part of the site, notably the immense Tempio di Giove Olimpico, the size of a football pitch. Here, too, is the marvellous Garden of Kolymbetra. Lovingly restored, this is a fine example of an Islamic garden and a lush oasis to rest mind and body after a morning's temple-gazing.

But let's not forget the Greeks made their mark across Sicily. About 100 km (62 miles) west of Agrigento, the Greek temples of Selinunte are strewn over a ridge directly above the sweeping sands of Marinella, and are one of the finest sights on the island. The ancient colony, named Selinous in ancient Greek after the wild celery *(selinon)* that still grows in abundance here today, reached the peak of its power in the 5th century BC. Its vast fertile lands made it a tempting target, however, and in 409 BC it was sacked by Carthaginians from across the sea in North Africa. Sixteen thousand people were killed and 5,000 taken as slaves.

Despite the tragic history, the ruins are awe-inspiring. There are seven temples, handily named A–G. The most complete are Temple E, possibly dedicated to Hera, and Temple C, with inscriptions to Apollo, which occupies the highest point of

Top The ruins of the Tempio di Giunone at Agrigento ***Above*** *The remains of the Tempio di Castore e Polluce at Agrigento, framed by a blossoming almond tree*

the Acropolis. The site is vast and never feels crowded. One of the real joys is that it is a wonderful place to walk, especially in spring, simply admiring the ruins of the temples and the views out to sea. An experience further removed from touring the Parthenon in Athens is hard to imagine.

Still visiting the Acropolis of Athens?
Yes, the Acropolis is impressive close-up, but for a wonderful (and relatively peaceful) view of it, climb to the top of Philopappos Hill a little to the southwest. Bring a picnic and watch the sun set behind the magnificent monuments.

Getting There *Most international flights to Sicily arrive at Palermo's Falcone Borsellino Airport. Trapani-Birgi Airport caters to several low-budget airlines.*

www.valleyofthetemples.com; www.selinunte.net

Chichen Itzá draws many visitors to its well-preserved Mayan pyramids and ball courts. The lost city of El Mirador, on the other hand? Enveloped in silent jungle, it's reached by the intrepid few.

the alternative to Chichen Itzá, Mexico

EL MIRADOR

Guatemala

The city of El Mirador was built over 2,000 years ago by Mayans living in the Mirador Basin in northern Guatemala, and is one of the oldest known Mayan sites in the world. It was once a vibrant metropolis covering about 36 sq km (14 sq miles); today, its ruins lie in ceremonious silence, swathed deep in the Petén Jungle. Unlike Mexico's Chichen Itzá, El Mirador is untouched by the cacophony of souvenir vendors and mass tourism. Only the most adventurous need apply.

Located 40 km (25 miles) from the nearest paved road, the lost city of El Mirador is typically reached on a two-day guided hike from the village of Carmelita in dry season (December to April). Various outfitters run tours, but we recommend booking directly through the Carmelita community, whose trained and licensed guides are part of a sustainable tourism initiative contributing to the livelihood of locals. As you snake along the jungle trails, you'll spy monkeys and toucans galore and, if you're lucky, even a puma or jaguar. Magnificent El Mirador is dominated by two colossal pyramid complexes and, unlike at Chichen Itzá, you can actually climb up them. La Danta rises to a height of 70 m (230 ft) from a hillside, making it the world's tallest Mayan pyramid. El Tigre is no less impressive for its huge base. As you gaze out at this remote ancient city from the heights, the softly humming jungle will be the only sound you'll hear.

Still itching to go to Chichen Itzá?

Avoid visiting on a Sunday, when the site tends to be busiest since entry is free for all Mexican nationals.

Getting There *Fly from La Aurora Guatemala International Airport to Flores. Buses run from Flores to Carmelita.*

www.turismocooperativacarmelita.com

La Danta pyramid rising out of the thick foliage of the Petén Jungle

There are no city ruins more evocative of millennia gone by than those of Pompeii. But the ancient city of Kourion in Cyprus has incredibly well-preserved floor mosaics that are no less magnificent.

the alternative to *Pompeii, Italy*

KOURION

Cyprus

A beautiful 4th-century floor mosaic in the House of Eustolios

Perched above Cyprus's glittering blue Episkopi Bay, the ruins of Kourion echo the city's heyday as a jewel in the crown of Rome's far-flung empire.

Like Pompeii, Kourion was an ancient city humbled by natural forces: a succession of earthquakes in AD 5 destroyed many of its buildings and drove away its inhabitants. Despite this catastrophe, the site today offers an irresistible glimpse into Kourion's past, from the mighty amphitheatre, where thousands once cheered gladiatorial combats, to private villas with beautiful mosaic floors.

These mosaic floors, just like those at Pompeii, are remarkably well preserved. Within the roofless walls of the House of Eustolios, polychrome tiles depict mythical figures, doves, guinea fowl and magpies. Faded tesserae paving the House of the Gladiators portray slave-warriors armed for mortal combat in the arena. The luxury of those who lived here is evident.

The pillars of the Roman agora, the commercial heart of the city, and the Nymphaeum, a communal bathing place and social hub, provide more echoes of everyday life in the ancient world. So too does the House of the Earthquake: here archaeologists found the skeletons of a young family, clinging to each other in their final moments – a teenage girl and three older men. Like those overwhelmed by the eruption that tore down Pompeii, they too fell victim to the cataclysm that wrecked their home. It might be ancient history, but the stories of the past have never felt more real.

Still heading to Pompeii?

To get the most out of your visit, join a tour or take a detailed guidebook. The surfaces are uneven so wear comfortable shoes.

Getting There *The nearest international airports to Kourion are at Paphos (about a 45-minute drive) and Larnaca (around one hour).*

www.visitcyprus.com

The world's loftiest brick tower, Delhi's Qutb Minar is more than a match for its Pisan leaning cousin. The supremely ornate carvings make this soaring turret a masterpiece of Islamic architecture.

the alternative to
the Leaning Tower of Pisa, Italy

QUTB MINAR

India

A soaring cylindrical tower of ornately carved red sandstone and white marble, the magnificent Qutb Minar in Delhi is the world's tallest brick minaret. At 73 m (240 ft), the tower surpasses its 57-m (180-ft) Italian contemporary in Pisa. If you look closely, you might spot that it's slightly askew from the vertical, though not as alarmingly so as its Pisan peer.

Instead of a tourist-pleasing tilt, it's the elaborate carvings, etched into the glowing-red sandstone bands, that make the Qutb Minar a showstopper. Encrusted with gorgeous arabesque patterns and calligraphy bearing Qur'anic verses, it's one of the first Islamic structures in India. The first storey was built in 1193 under Sultan Qutb-ud-din to celebrate victory over the Hindu kingdom. Successive sultans added more storeys until the tower was finally completed in the 14th century, which perhaps explains why it slants a little.

Part of a complex containing a 12th-century mosque and even older funerary tombs, all surrounded by peaceful gardens, the tower isn't the only stunning sight here. Even so, it's not as if you need more reasons to visit: the staggering Qutb Minar is more than enough.

Still heading to Pisa's Leaning Tower?
If you don't want a long wait for one of the limited places on a guided walk up the tower, book a ticket well in advance.

Getting There *Buses from Delhi's Indira Gandhi International Airport take about 40 minutes to reach the Mahrauli suburb, home to the Qutb Minar.*

www.delhitourism.gov.in

The ornate Qutb Minar, towering above a mosque and funerary tombs

While the photogenic ruins of Angkor Wat have been splashed all over the big screen, Thailand's atmospheric Sukhothai Historical Park remains far less familiar.

Phra Achana, the largest Buddha statue in Sukhothai, inside the roofless Wat Si Chum

the alternative to *Angkor Wat, Cambodia*

SUKHOTHAI HISTORICAL PARK

Thailand

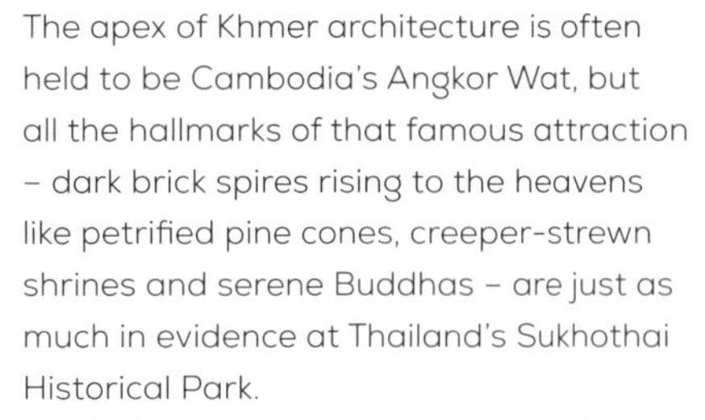

The apex of Khmer architecture is often held to be Cambodia's Angkor Wat, but all the hallmarks of that famous attraction – dark brick spires rising to the heavens like petrified pine cones, creeper-strewn shrines and serene Buddhas – are just as much in evidence at Thailand's Sukhothai Historical Park.

Sukhothai is both a portrait of elegant decay and an architectural photo album: a gallery of styles left behind by the succession of empires who occupied this once great city. The strongest echoes of

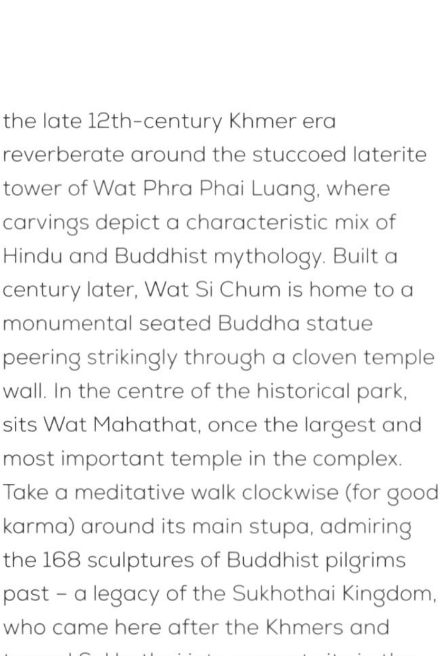

the late 12th-century Khmer era reverberate around the stuccoed laterite tower of Wat Phra Phai Luang, where carvings depict a characteristic mix of Hindu and Buddhist mythology. Built a century later, Wat Si Chum is home to a monumental seated Buddha statue peering strikingly through a cloven temple wall. In the centre of the historical park, sits Wat Mahathat, once the largest and most important temple in the complex. Take a meditative walk clockwise (for good karma) around its main stupa, admiring the 168 sculptures of Buddhist pilgrims past – a legacy of the Sukhothai Kingdom, who came here after the Khmers and turned Sukhothai into a great city in the 13th century. When their successors,

Bell-shapaed chedis, which are dotted all over the grounds of Sukhothai Historical Park

the Ayutthaya, abandoned Sukhothai in the 16th century, they left some 400 of the bell-shaped chedis (Thai stupas) scattered among the grounds.

Sukhothai translates as "dawn of happiness". Visit in the small hours as the sun begins to blush the cheeks of the great Buddha pink and gold, and you'll be hard pressed to think of a better name.

Still going to Angkor Wat?
Visit in off-season between June and October for a slightly quieter experience. The temples cover a vast area, so be prepared to set aside at least three days for exploring the site.

Getting There *It's a 40-minute taxi ride from Sukhothai Airport to Sukhothai Historical Park.*

www.tourismthailand.org

Beneath the wide-open skies of Arizona's desert sit the ancient cliff dwellings of Canyon de Chelly – and they're just as awe-inspiring as the more famous Mesa Verde in Colorado.

the alternative to *Mesa Verde, USA*

CANYON DE CHELLY

USA

Tucked into a remote corner of Arizona, the soaring red-rock buttes and parched desert backdrop of Canyon de Chelly contain hundreds of dramatic cliff dwellings built into the caves and recesses of the canyon walls. Constructed nearly 1,000 years ago, they were abandoned in the 13th century, the time when the first dwellings at Colorado's Mesa Verde were carved. You'll spy pictographs as you descend to the canyon floor to reach the most striking dwelling, the White House. To see other ancient homes, take a tour led by Navajo guides, who'll give you a deeper insight into the former Ancestral Puebloan peoples who once thrived in this seemingly barren landscape.

Still want to experience Mesa Verde?
The Wetherill Mesa in the western section draws the fewest visitors year-round.

Getting There *The nearest airport to Canyon de Chelly is in Flagstaff, Arizona. From here it's a three-hour drive east to the canyon.*

www.nps.gov/cach

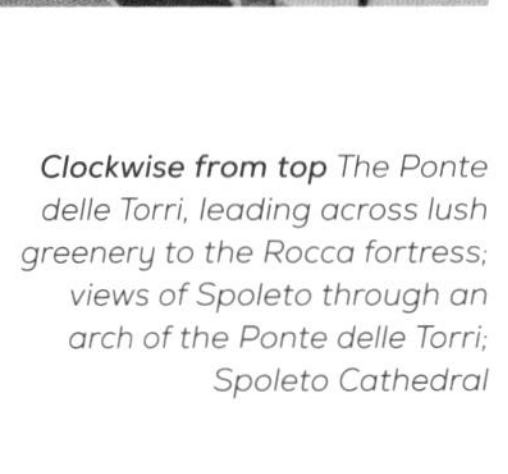

***Clockwise from top** The Ponte delle Torri, leading across lush greenery to the Rocca fortress; views of Spoleto through an arch of the Ponte delle Torri; Spoleto Cathedral*

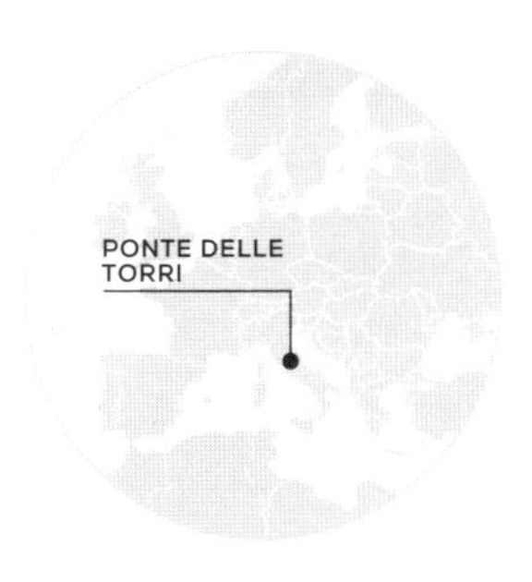

While Pont du Gard is the world's most famous aqueduct, its only companions are a visitor centre and car park. The equally breathtaking Ponte delle Torri in central Italy, however, leads you into the heart of the medieval town of Spoleto.

the alternative to *Pont du Gard, France*

PONTE DELLE TORRI

Italy

Emerging from the Mediterranean scrub near Nîmes in southern France, the ancient Roman Pont du Gard aqueduct runs for a heroic 140 m (459 ft) before disappearing again into the bush. Impressive though it is, after walking its length, there's not much left to do but head back to the coach-filled car park. If you're after more, consider, instead, the Ponte delle Torri in Umbria, an equally awesome acqueduct with added bonuses.

No one is sure exactly when the Ponte delle Torri was built, but it's thought to date to about the 14th century. It brought water from the mountains to the upper part of Spoleto, and served another purpose, too: bookending the 230-m (755-ft) span are two castles, the fearsome Rocca, a medieval fortress, and the Mulini fort. The bridge itself is certainly an impressive sight, and so too are the views from it. The thickly wooded Tessino Valley extends in both directions while the silent Apennines rear ahead.

After pausing to take it all in, it's mere minutes to the centre of Spoleto itself, a town of narrow streets, early Christian basilicas and an impossibly lovely cathedral. Spoleto is named one of Italy's most beautiful towns but it's the aqueduct that will draw you to return. Contemplating the rugged scenery, it seems hard to believe that a thriving medieval town lies just metres from your back.

Still want to see Pont du Gard?
For the best way to see the bridge, take to the water. You can either swim beneath it or hire a canoe from nearby Collias.

Getting There *Perugia, 50 km (30 miles) from Spoleto, is the nearest airport. Rome is a 1.5-hour train journey.*

www.umbriatourism.it

FESTIVALS AND PARTIES

Streets thrumming with theatrical parades, joyous dancing and pounding drumming. No, this isn't Rio, it's the carnival in Montevideo. Skip the world's largest street party for its smaller cousin in Uruguay and you're guaranteed just as much fun.

the alternative to *Rio Carnival, Brazil*

MONTEVIDEO CARNIVAL

Uruguay

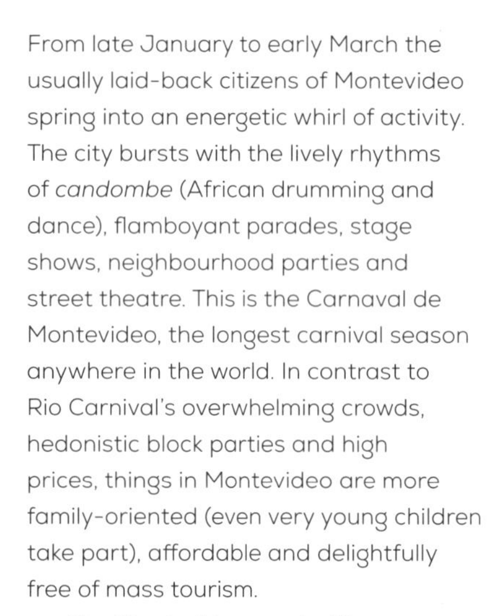

From late January to early March the usually laid-back citizens of Montevideo spring into an energetic whirl of activity. The city bursts with the lively rhythms of *candombe* (African drumming and dance), flamboyant parades, stage shows, neighbourhood parties and street theatre. This is the Carnaval de Montevideo, the longest carnival season anywhere in the world. In contrast to Rio Carnival's overwhelming crowds, hedonistic block parties and high prices, things in Montevideo are more family-oriented (even very young children take part), affordable and delightfully free of mass tourism.

The Montevideo carnival is more than anything a celebration of Afro-Uruguayan culture. The rhythms and soul of *candombe* and tango, grown from African roots and infused with a Uruguayan flavour through the years, are very much at its heart. *Comparsas* (groups of *candombe* drummers, dancers and flag-bearers) practise all year round for the parades. And all that preparation culminates in the most important parade of all, the Desfile de Llamadas (The Calls Parade), named after the calls of the thundering drums summoning people to the party. Over two days in early February, wave after wave of *comparsas* marches through neighbourhoods, the overwhelming beating of drums luring people into the streets. Thousands of dancers in bright tail feathers pull bystanders to join in the parade. The street party is now in full swing.

In between the major parades, *murga* shows (theatre featuring drumming, singing, dancing and mime) take place in the Teatro de Verano, an open-air venue in Parque Rodó, and on *tablados* (temporary stages) dotted around the city. Each neighbourhood has its own

Clockwise from left Candombe dancers and drummers calling people to the party in the Desfile de Llamadas; costumed dancers enacting folktales

troupe that puts on mainly satirical performances that poke fun at modern-day life and appeal to both locals and visitors, regardless of their Spanish-speaking abilities. With plenty of productions for kids too, these shows highlight the inclusivity of Montevideo's celebrations. So, whether you're here for a 40-day whirlwind of a party with friends, or you're travelling with the family, you're guaranteed a carnival to remember.

Still partying at Rio Carnival?

Book accommodation well in advance (most places impose a minimum stay of four nights). Leave your valuables in the safe in your hotel while you're out pounding the pavements and above all else, remember to hydrate: partying in Brazil's sweltering summer is sweaty work.

Getting There *Montevideo is served by Carrasco International Airport. Buses run to the city centre, taking about one hour.*

www.descubrimontevideo.uy/en/carnival

The world's biggest arts festival, Edinburgh Festival Fringe may get all the attention, but on the other side of the planet, the Wellington Fringe is quietly making a name for itself.

the alternative to *Edinburgh Festival Fringe, UK*

WELLINGTON FRINGE

New Zealand

In the crook of a finger of land at the very south of the North Island, Pōneke (the Māori word for New Zealand's capital city) is backed by green hills, with the azure Cook Strait to the south and the rolling Remutaka Range to the east. Every year for three weeks in late summer, from mid-February to mid-March, the Wellington Fringe takes over the city, entertaining with the best and brightest homegrown talent and plenty of *manaakitanga* (hospitality).

Around 700 shows are hosted in places all across the city, from the entertainment district Te Aro to the arty suburb of Mount Cook. Performances are just as likely to be found in the boat-like interior of Old St Paul's Cathedral, in bars, squares, street corners, parks and hotel foyers, as in the intimate BATS Theatre (the original 1990s home of the Fringe) and Te Auaha, the city's premier creative venue.

So what can you expect? "Wellywood" is no stranger to celebrity. Past festival acts have included New Zealand comedy duo Flight of the Conchords, filmmaker and actor Taika Waititi and Russian punk rock group Pussy Riot. But it's very likely that you'll be experiencing a performer new to you and everyone else in the audience. Dubbed "New Zealand's birthplace of brilliance", the

The wooden interior of Old St Paul's Cathedral, resembling the upturned hull of a galleon

Wellington city centre and harbour on Cook Strait, surrounded by verdant hills

Wellington Fringe is dedicated to nurturing up-and-coming Kiwi talent. One night you'll be snorting with laughter at a piece of comedy improv from the city's latest funny woman, the next gasping in delight at a high-energy circus act.

Like Edinburgh's Fringe, Wellington's is vibrant, fun and sometimes downright odd. Bring an open mind and get ready to be entertained.

Still heading to Edinburgh Festival Fringe?

Reserve your accommodation as far in advance as you can (some places are booked up a year ahead). Leave the car at home and, for a peaceful night, consider staying in West Edinburgh close to a tram stop – it's no more than a 20-minute ride into the centre even from Edinburgh Park (which is close to the airport).

Getting There *Wellington International Airport is 8 km (5 miles) south of the city. Frequent Metlink public buses run to the city centre.*

www.fringe.co.nz

BALI ARTS FESTIVAL
Indonesia

Known locally as Pesta Kesenian Bali (PKB), this arts festival in Bali's capital, Denpasar, takes place each year from mid-June to mid-July. Music, dance, theatre and art and craft exhibitions are held in the Taman Werdhi Budaya Arts Centre.

AVIGNON FESTIVAL
France

Founded in 1947, the Festival d'Avignon celebrates all aspects of culture over three weeks in July, and is the oldest festival in France. All the venues are open-air and include the inner courtyard of the splendid Gothic Palais des Papes.

BRIGHTON FRINGE
UK

The largest annual arts festival in England takes place in May and June each year in more than 160 venues across Brighton. With music, dance, theatre, circus and more, this festival focuses on local talent, and anyone can take part.

The tomato-throwing frenzy of La Tomatina, held in Buñol since 1945, may be the world's most famous food fight, but there's another quirky festival in Haro. With an even older tradition, begun in 1290, the Wine Battle promises to drench you in red wine. Thirsty yet?

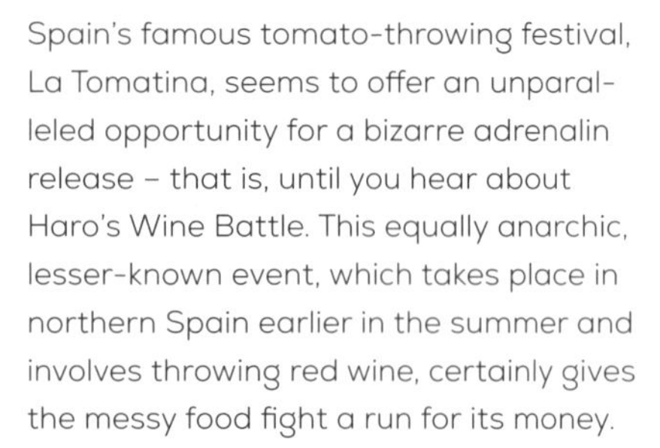

the alternative to La Tomatina, Spain

HARO WINE BATTLE

Spain

Spain's famous tomato-throwing festival, La Tomatina, seems to offer an unparalleled opportunity for a bizarre adrenalin release – that is, until you hear about Haro's Wine Battle. This equally anarchic, lesser-known event, which takes place in northern Spain earlier in the summer and involves throwing red wine, certainly gives the messy food fight a run for its money.

Unlike La Tomatina, which is a fairly recent invention, the Haro Wine Battle has a long history behind it. According to a royal proclamation of 1290, the small wine-producing town of Haro had to reassert, annually, its right of possession over the local vineyards (where the battle now takes place), against a claim by the nearby town of Miranda del Ebro. The event began as a low-key bounds-beating ceremony that involved hoisting a flag and holding Mass. In 1906, however, gleeful participants decided to liven proceedings up by "baptizing" each other with wine. Inevitably, it wasn't long before this ritual became a chaotic wine battle that overshadowed the event's more sober formalities.

Today, the pattern of the battle is well established. It begins with a procession of around 5,000 people, wearing white T-shirts and clutching pieces of religious paraphernalia, including crucifixes and bibles. They walk from Haro to the site of an isolated chapel, where a flag is raised and a Mass said. Once this ceremony has ended, battle commences. In a matter of minutes, all 5,000 or so pristine white T-shirts are stained a deep shade of purple from over 20,000 litres of wine flung into the air. Anyone nearby is fair game – including photographers and television camera crews, who come prepared with protective plastic (this isn't their first rodeo). Water pistols, *botas* (leather drinking bottles), plastic bottles and buckets are the common weapons

Haro Wine Battle participants spraying each other with wine among the vineyards

Top A tractor carrying a, mostly dry, crowd to the site of the chapel for the beginning of the battle
Above *The aftermath: partying revellers drenched in red wine*

of war, but experienced participants who really mean business come equipped with ultra-modern industrial crop-sprayers fed by tanks on their backs.

When supplies of ammunition have been exhausted, the berry-coloured crowd troops down the hill to feast on snails and red wine. They then head back, still in procession, to Haro – drying off in the sunshine on the way. Back in town, the party continues with singing, dancing, fireworks, and, of course, a glass of wine.

Still want to go to La Tomatina?

The crowds at the tomato-throwing free-for-all in Buñol are simply unavoidable. If you're keen to join the revelry, swot up on the rules before you arrive: squash all your tomatoes before you throw them, keep a safe distance from the lorries and don't throw any hard objects.

Getting There *Haro's nearest international airports are at Vitoria and Bilbao. The town also has bus and rail connections with Bilbao and Barcelona.*

www.haroturismo.org

An unusual twist underpins the St Patrick's Day celebrations in Montserrat. Forget Dublin and head to this Caribbean Emerald Isle for a uniquely special party you won't forget.

the alternative to
St Patrick's Day, Ireland

ST PATRICK'S DAY

Montserrat

Typical masks worn by Montserratians during St Patrick's Day celebrations

The small Caribbean island of Montserrat is one of only a few places in the world besides Ireland where St Patrick's Day is a national holiday. But these are not your typical celebrations; they combine a unique African and Irish heritage.

It all goes back to a significant moment in the island's history. On 17 March 1768, Montserrat's enslaved people organized a rebellion against the Irish settlers who were gathering to celebrate St Patrick's Day. The revolt was quashed and nine enslaved people were killed. More than two-and-a-half centuries later, today's Montserratians commemorate this event with their take on St Patrick's Day celebrations to honour the uprising and remember their shared history.

Ever since becoming a national holiday in 1985, the festivities in Montserrat have stretched over ten days, which is double the five in Dublin. In a salty sea breeze, masked dancers sway to Caribbean rhythms, street stalls sell finger-licking local food and families participate in a "freedom run" and play traditional games. There's even a fashion show of costumes made from Montserrat *madras*, the island's national textile. Undeniably green, the fabric may be the only truly predictable element of this St Patrick's Day celebration.

Still heading to Dublin for St Patrick's Day?
Join the Harbour2Harbour charity walk, which supports local organizations. It takes place along beautiful Dublin Bay on or just before 17 March.

Getting There *The nearest airport to Montserrat is in neighbouring Antigua. From there it's a 20-minute flight or a 90-minute ferry ride.*

www.visitmontserrat.com

Bavarian songs, chequered dirndl, piles of pretzels: welcome to Straubing's Gäubodenvolksfest, Bavaria's second-largest folk festival. It's a more local affair and just as much fun as its bigger relation in Munich.

the alternative to *Oktoberfest, Germany*

GÄUBODENVOLKSFEST

Germany

From spring to autumn, Bavaria comes alive with folk festivals. While Munich's Oktoberfest has gained international fame, Straubinger Volksfest's similar merrymaking flies under the radar.

In the town of Straubing, about 145 km (90 miles) from Munich, the Gäubodenvolksfest takes place every August. Dating back to 1812, it's only two years younger than its more famous big cousin and the similarities don't end there. Here too you'll find buzzing tents serving up huge glasses of beers, fluffy candy floss, gaudy fairground rides and much jollity, not to mention lederhosen and dirndl. In fact, as the crowd is mainly local, you're likely to see fewer outfits sourced between the airport and the hotel, and more of the real thing. These are truly traditional festivities, full of revellers tucking into plates piled high with roast chicken and potato salad, while bands play cheery folk tunes. All around is the hum of excited chatter – you'll hear a lot of Bavarian dialect and German. So, don traditional attire, learn a few refrains and join the conviviality: prost!

One of the seven large beer tents full of people feasting, drinking and listening to music during the Gäubodenvolksfest

Still going to Oktoberfest?

The Oktoberfest is busiest at weekends and in the evenings, so try to visit Monday to Thursday. Dress the part: various outfitters around Munich rent out Bavarian folk costumes.

Getting There *Trains from Munich International Airport take one hour and 50 minutes to reach Straubing.*

www.straubing.de

A boutique festival with an extravagant attitude, Wilderness is the sequin-covered alternative to Glastonbury, packing musical mayhem and immersive escapism into a quintessentially British weekend.

the alternative to *Glastonbury Festival, UK*

WILDERNESS FESTIVAL

UK

Every August, Wilderness Festival rolls into the ancient woodlands of Oxfordshire. And with it? Just 10,000 guests, a bunch of talented musicians and an indulgent number of gourmet food trucks. This community-focused festival offers all of Glastonbury's musical magic, with plenty of extras thrown in for good measure.

The first thing you'll notice when you arrive here are the costumes. Wilderness is big on fancy dress, providing costume themes, like "What Comes Naturally" and "Maximalism", for each day of the festival. These themes are pretty broad, leaving festival-goers ample room to get creative. And creative they get. Expect to see a dazzling array of sparkly spandex, colourful hats, biodegradable body glitter and a cape or two, whatever the weather.

Once you've sorted your outfits, it's time for some music. Wilderness doesn't subscribe to one specific musical genre. Rather, this is a festival for everyone, with six different venues showcasing a range of entertainment for young families and party-loving friends alike.

Crowds watching Björk, one of the headliners at Wilderness Festival in 2015

***Above** Festival-goers practising paddleboard yoga on Wilderness's scenic lake*
***Right** Enjoying a fine meal in the Dining Room Tent*

Here, folk bands, hip-hop artists and dance acts share stages with brass bands, full-blown orchestras and chart-toppers like Grace Jones and the Flaming Lips – and there are performances everywhere you look. This quirky festival isn't all about music, though. Literary tents offer spoken-word shows, poetry slams and book readings, while immersive theatre troupes march through the fields, recruiting participants as they go.

Then there are the activities. Friends challenge one another at axe-throwing competitions or test their balance with paddleboard yoga. Those in need of some peace and quiet often retreat to the shade for spoon carving or life drawing sessions with like-minded strangers. There's also wild swimming in the sun-dappled lake if you're prepared to brave the cold.

There's a level of indulgence and pampering available too, giving Wilderness its boutiquey reputation. Festival-goers can relax with a massage or warm up in one of the wood-fired hot tubs by the lake – a far cry from Glastonbury's muddy fields and wet wipe showers. You can even wield a bat at the annual Sunday cricket match: just keep an eye out for the "surprise" streaker (it happens every year).

Foodies will have a field day here as well. Banquet tables laden with treats prepared by award-winning chefs can be pre-booked for an evening meal. The gourmet food trucks are just as good, serving up bao buns and sushi burritos to the costumed masses.

After a day of fun and food, late-night ravers head for the laser discos in the woods or gather around jazz musicians leaping about the carousel stage. Meanwhile, flower-garlanded figures lie back on the grass to count the stars before

Top A round of crazy golf
Above Soaking in the toasty hot tubs by the side of the lake

heading back to their glamping cabins. After four days, you'll emerge from a hidden world. This place feels extravagant and joyous, providing a sense of exhilarated fun many haven't felt since childhood. Wilderness is perhaps a microcosm of Glastonbury in its purest form: a space for embracing the joy of connecting with strangers. There's always a campfire burning and always an invitation to join those around it.

Still going to Glastonbury?

Glastonbury Festival welcomes a whopping 200,000 visitors over five days. If you're lucky enough to nab a ticket, remember to pack your sturdiest walking shoes, buy a brightly coloured tent and make a note of where you've parked your car.

Getting There *Charlbury is the closest train station, with regular connections to London Paddington.*

www.wildernessfestival.com

A playground for jazz titans past and present, Copenhagen's sprawling festival reverberates through the city, eclipsing its more famous, and more compact, Swiss counterpart.

the alternative to *Montreux Jazz Festival, Switzerland*

COPENHAGEN JAZZ FESTIVAL

Denmark

As the city's cool Scandi temperatures begin their rise towards summer highs there tends to be one classical music event on everyone's lips: Copenhagen Jazz Festival. Scandinavia's largest jazz celebration is ripe with history and offers up just as good a musical feast as Switzerland's Montreux Jazz Festival.

Jazz music has echoed through Copenhagen's historic streets since the 1950s and 60s, when the city became a hub for icons such as Ben Webster, Dexter Gordon and Thad Jones. The jazz festival itself dates back to 1964 (making it a few years older than Montreux's), when the Tivoli Gardens put on an event featuring the likes of Miles Davis and Thelonious Monk. Nowadays, it's heavyweights like Herbie Hancock that steal the show.

While Switzerland's jazz festival is concentrated on a number of big stages, often packing a 200,000-plus partisan crowd into a single venue, Copenhagen's July bonanza casts its berth over countless locations. From the revered bastions of Jazzhus Montmartre and La Fontaine to the city's small, hipster cafés, the festival welcomes aficionados all over the city. Further afield, venues such as the Copenhagen Opera House also offer an unparalleled opportunity to experience the genre in the finest of surroundings.

Inclusivity, however, is at the heart of Copenhagen's festival. Visitors are never far from a free show here, with pavement performances providing music for the masses. Everywhere you walk you'll hear the staccato sounds of contrabass thuds, dazzling trumpet solos and clapping crowds. As a festival that caters for all (children, retired locals and restless teens alike), the atmosphere here is special – derived

Picnicking at one of the numerous open-air concerts taking place during the festival

Danish-Finnish bass player Lennart Ginman and The Shape of Jazz performing a set at the Statens Museum

from an ever-changing cocktail of world-class music, memorable weather (good, great, torrential and everything in between) and a strong community of music-lovers.

For those visiting the city in February, there's also the Copenhagen Winter Jazz. Though smaller than the summer show, Winter Jazz places even more importance on promoting local volunteer groups and societies – the music's pretty great, too.

Still going to Montreux?

For an up-close-and-personal Montreux experience, give the two main stages a miss in favour of the smaller Montreux Jazz Cafe and the cluster of tiny open-air venues.

Getting There *The festival is held across the city centre, which is a ten-minute metro or train ride from Copenhagen Airport.*

www. jazz.dk

LUNAR NEW YEAR, MALAYSIA

Home to a large Chinese population, the island of Penang hosts an exhuberant Chinese New Year festival, which is just as eventful as Hong Kong's beloved celebration.

the alternative to
Lunar New Year, China

LUNAR NEW YEAR, MALAYSIA

Malaysia

Forget the busy street parades along Hong Kong's Nathan Road and think of the viridian flanks on Penang Hill. This is where the Malaysian island's 16-day-long Chinese New Year celebrations begin with a bang: bedazzling fireworks are lit over Kek Lok Si, one of the largest Buddhist temples in Southeast Asia. The grandest night of celebrations is the ninth, when events move to Penang's bustling city of George Town. Here, the promenade is taken over by tables laden with tropical fruits, religious icons and ceramic urns filled with wafting joss sticks. Lion dances on stilts and thunderous drumming keep the crowd lively till midnight, when the sky, once again, fills with exploding fireworks.

Still going to Hong Kong?

Book a restaurant table with a view in advance to enjoy the firework display over Victoria Harbour.

Getting There *Penang airport is connected to Malaysia's international hub Kuala Lumpur, which is 370 km (230 miles) away by road or express ETS train service.*

Weary of being lifted to heights of ecstasy by a BBC Prom, only to step out onto a hectic London street afterwards? Try Italy's Spoleto Festival, with concerts in a Roman theatre and a leisurely walk home along pretty medieval streets.

***the alternative to** the BBC Proms, UK*

THE FESTIVAL DI SPOLETO

Italy

Running for 16 days each summer, the Festival di Spoleto takes over the entire Umbrian hill town of Spoleto, and as a visitor you will live and breathe the festival from morning unto night. Of course, London's BBC Proms are unique – where else is there a music festival that lasts for three months? But wonderful as it is to be able to catch a Prom in the evening after a busy day's sightseeing, there is much to be said for the fully immersive experience of Spoleto. As you wander the picturesque cobbled medieval streets of the historic centre you might catch the sounds of an orchestra practising in the Roman theatre or even, if you are lucky, a world-famous contralto rehearsing in the cathedral.

This must-see festival for art and culture fans was founded in 1958 by Italian-American opera composer and playwright Gian Carlo Menotti. An anti-elitist, Menotti was determined to build on his conviction that "art was not just a liqueur at the dinners of the rich". He settled on the beautiful town of Spoleto, already blessed with several theatres and a supportive mayor, for the extravaganza he called La Festival dei Due Mondi (the "two worlds" being Europe and America). Not only did he use the festival to attract emerging talent, but he had the clout and charm to persuade international megastars to perform for a steal. Thus, the festival has attracted a glittering array of international musicians, actors, directors, writers and dancers over the years, ranging from filmmaker and auteur Luchino Visconti, who directed *Macbeth* for the opening night of the very first festival, to the shiny-pated TV star of *Inspector Montalbano*, Luca Zingaretti.

Clockwise from left Fireworks above the town at the end of the Festival di Spoleto; performers in action at the festival; musicians outside a café in the old town centre

Piazza del Duomo, the heart of the town and the setting for outdoor concerts during the Festival di Spoleto

Spoleto was soon one of the leading cult festivals of the 1960s, and became known as the Edinburgh of the South, drawing big-name theatre, music and dance performers, including Rudolf Nureyev, Luciano Pavarotti, Jacqueline Du Pré and Margot Fonteyn, to name but a few.

Today Spoleto might not be as famous on the international festivals scene as it was in the 1960s, but every year, without fail, its offerings form an irresistible 360-degree experience for ears, eyes, mind and soul. Running from the last week in June through the first ten days in July, Spoleto literally becomes its festival. The cafés and streets fill with the sounds of rehearsals and opinionated chatter about the previous nights' performances. The gilded boxes of the 19th-century Teatro Nuovo are polished to a gleam, the cathedral square is transformed into a stage, and everywhere are technicians lugging huge boxes of kit and sticky-taping cables to the cobbles. People-watch carefully, and you might spot the young ballerina you saw last night as Cinderella rushing up Via Apollinare, or the first violinist of the festival orchestra snatching a quick cigarette and espresso between rehearsals. Hairdressers are worked off their feet, fashion boutiques make a killing and in the evenings the streets become a perfumed catwalk for the immaculately dressed of all ages heading to the theatres.

What is equally wonderful about Spoleto is that you can easily escape the buzz at any moment. The town is surrounded by the glorious Umbrian hills, and there are any number of footpaths to take you to a place where the only sounds are made by birds. One of the best trails follows the Spoleto–Norcia disused railway line into the Valnerina valley. This includes stretches through a few tunnels (so make sure your phone has a torch).

And if you fancy exploring more of Umbria's cultural attractions, Spoleto

Human puppets from a street theatre company performing during the festival

A performance of classical ballet, which forms a big part of the Festival di Spoleto's programme

is a great base for seeing the region's other fabulous towns. Train and bus services are good, so you don't even need a car to visit such jewels as Orvieto, Perugia, Todi and Assisi, all of which are crammed with beautiful art that eclipses that in most of the UK's home counties surrounding London.

Just two words of warning: Spoleto is hugely popular with Italians during festival time, so book accommodation well in advance. Restaurants can be pricey and pre-booked too, and you may decide that self-catering is the best option, to say nothing of being a fabulous excuse to shop in Spoleto's morning market and irresistible delis.

Still going to the BBC Proms?

Seats for the most popular concerts sell out fast but a number of standing tickets are released for purchase on the day. If you plan to watch a lot of performances, invest in a season ticket. Tickets for the Last Night of the Proms are notoriously hard to come by, but you can instead book for the concert in Hyde Park on the same night. This open-air extravaganza ends with a spectacular fireworks display.

Getting There *The nearest airport is Perugia, 48 km (30 miles) from Spoleto. Shuttle buses run between the train station and the historic centre.*

www.festivaldispoleto.com

While Coachella is known for its brand-sponsored events and famous guests, Bonnaroo – a world-class festival in rural Tennessee – is all about the communal spirit and good vibes. It just might be the closest you get to Woodstock this side of 1969.

***the alternative to** Coachella, USA*

BONNAROO

USA

Coachella may be the hottest music ticket in town, but for those looking for a fun-loving festival that doesn't take itself too seriously, Bonnaroo is one for the books.

Every June, a large farm in the sleepy town of Manchester, Tennessee turns into a candy-coloured metropolis of art and music, welcoming thousands for four days of peace, love and music. With roots as a jam-band festival – but now pulling acts of all genres, from the Foo Fighters to Lizzo – Bonnaroo comes with a sense of freedom that's less about the label on your festival garb and more about the energy you bring. The beloved 'Roo mantra is "Radiate Positivity", and that's what you'll find here: strangers are quick to make friends with a "hi" and a high five.

Community spirit and sheer fun abound here. During the day, festival-goers, slick from the southern heat, race from stage to stage to catch their next favourite band; others dance under the experimental art installations or retreat to the comedy and cinema tents for a sweet taste of air con. At night, discos move the party to the campground where most attendees stay, embracing all the mud and mayhem that comes along with it. A hotel for the night? That's just not Bonnaroo's style.

Still want to experience Coachella?

If the California desert is more your speed, pass on the expensive hotel and camp in the valley for a fully immersive experience.

Getting There *Nashville International Airport is the closest airport. You can rent a car here for the hour's drive to the farm.*

www.bonnaroo.com

Festival-goers enjoying a music performance at Bonnaroo, Tennessee

Venice's beloved Biennale may be vast, but what if you could see great art for free? International, eclectic and charged with a rebellious spirit, the São Paulo Biennial puts on provocative contemporary art without, amazingly, admission fees.

the alternative to *the Venice Biennale, Italy*

SÃO PAULO BIENNIAL

Brazil

Like its Venetian predecessor, the São Paulo Biennial showcases groundbreaking art, design and architecture from all around the world. The bonus with Brazil's version: you don't have to pay to see it.

Inaugurated in 1957, the biennial is the world's second oldest and now one of South America's most important art events. Its exhibitions, running from September to December, fill Ibirapuera Park, a lush oasis in the middle of the city's concrete jungle, as well as countless small galleries. Dozens of nations are represented here but the highlight is always the current crop of South American art. Peppered with sharp social and political insights, these exhibitions, which have included an examination of the slave trade and reflections on the Amazon from the perspective of an Indigenous artist, are never afraid to take a stand.

Still going to the Venice Bienniale?
Save a few euros and get early-bird tickets in March. Aim to visit between October and November when crowds start to wane.

Visitors passing a photography exhibition at the São Paulo Biennial

Getting There *São Paulo has a large international airport. The city's metro service is the most efficient way to travel between biennial exhibitions.*

www.bienal.org.br

MORE LIKE THIS

SHARJAH BIENNIAL UAE
In addition to art, music and film, the UAE's Sharjah Biennial incorporates substantial programming for families. Catch it on any odd-numbered year.

LIVERPOOL BIENNIAL UK
This biennial transforms the city with unusual and unexpected art in public spaces. Founded in 1988, it's the UK's largest art showcase and always promises to shake things up.

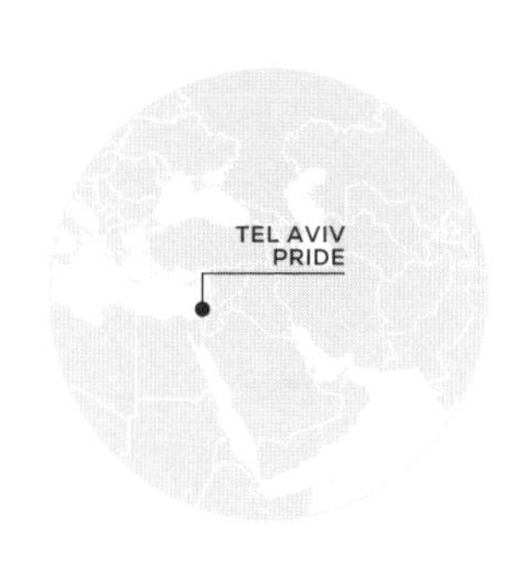

Hot on the heels of Sydney's huge queer event, Tel Aviv Pride – the largest such celebration in the Middle East – is just as exuberant, raucous and gleefully down to earth.

the alternative to *Sydney Gay and Lesbian Mardi Gras, Australia*

TEL AVIV PRIDE

Israel

In the run-up to Pride, city workers deck Tel Aviv's main boulevards in rainbow flags and posters announcing a smorgasbord of festivities. Each gloriously warm and sunny June, Tel Aviv Pride brings a variety of events – concerts, beach parties, art exhibits, film screenings, activities for teens and a drag festival – to the heart of Israel's high-tech metropolis. The energy, as expected, is always palpable.

It's clear why the Sydney Gay and Lesbian Mardi Gras is legendary – it's one of the largest such events in the world, with cabaret and comedy, drag and dance bringing colour (and thousands of tourists) to the city every year. Yet there's a lot to be said for choosing to celebrate Pride in Israel. Since it started in 1979, Tel Aviv Pride's strong visibility and unapologetic acceptance of diversity has helped stimulate debate about LGBTQ+ rights even in profoundly conservative and religious corners of Israeli society. Tel Aviv Pride is a huge party, of course, but – like in Sydney – it's also part of an enduring struggle for acceptance and equality.

We repeat: it's a party, and a great one at that. You'll hear the floats before you see them, a pounding techno beat heralding the arrival of each trailer carrying a DJ and dancers. As the parade makes its way along the beachfront promenade, bringing a Rio vibe to the Mediterranean, crowds from all walks of life – some pushing prams, others wearing outrageous wigs, many with both – join in the merriment, bouncing to the beat, celebrating LGBTQ+ rights and the joy of being alive.

Revellers parading through Tel Aviv, waving the traditional rainbow-coloured flags and garlands

Left *A dancer basking in the sun while masses party on the beachfront* **Below** *Moving through the streets in colourful festival dress*

Naturally, the highlight of the week-long extravaganza is the flagship parade, held on a Friday afternoon. Tens of thousands of tourists join over 200,000 Israelis in clubs, parks and, as in Sydney, on the fine-sand beach to celebrate a message of joy and love. The musical procession traditionally ends at Charles Clore Park, just north of the ancient old city of Jaffa, an ideal spot to lounge while watching the sun head towards the horizon.

Still set on partying at Sydney Gay and Lesbian Mardi Gras?

The huge Sydney Gay and Lesbian Mardi Gras festival brings a brash and flamboyant party to Australia's largest city every February and March. It's a joy to be a part of, so don a glittery body suit, pink tutu or whatever takes your fancy, and immerse yourself in the celebrations.

Getting There *Ben-Gurion International Airport (TLV), about 15 km (9 miles) southeast of Tel Aviv, is linked to the city by commuter rail and taxis.*

www.visit.tel-aviv.gov.il

MORE LIKE THIS

REYKJAVIK PRIDE
Iceland
A third of Iceland's population turns out for Reykjavik's Pride event, which brings six days of concerts, lectures, gallery openings, dances and a giant parade to the city in August.

PUERTO VALLARTA PRIDE
Mexico
The open-air concerts, beach parties, and revelry in nightlife venues and around town last a week in one of Mexico's premier LGBTQ+ hotspots.

The Holy Week celebrations in the ancient Inca capital of Cusco are just as exhilarating as Seville's grand processions. What is more, they mix ancient Andean and Catholic customs.

Traditional dancing in Plaza de Armas, which forms part of the festivities during Semana Santa

the alternative to
Semana Santa, Seville, Spain

SEMANA SANTA, CUSCO

Peru

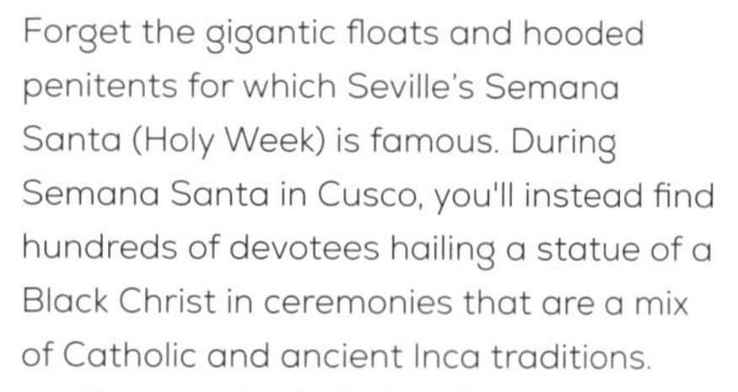

Forget the gigantic floats and hooded penitents for which Seville's Semana Santa (Holy Week) is famous. During Semana Santa in Cusco, you'll instead find hundreds of devotees hailing a statue of a Black Christ in ceremonies that are a mix of Catholic and ancient Inca traditions.

Thousands of pilgrims from mountain villages flock to Cusco for the week-long processions, festivities and re-enactments from the Bible. Stalls sell empanadas and corn bread on Plaza de Armas, and on Holy Thursday, people gather to share 12 dishes (representing the 12 apostles), of potatoes, seafood and *tarwi* (lupin beans).

Unlike Seville, where Good Friday is the focus of the festival, Holy Monday is the highlight in Cusco. At dawn in the city's cathedral, which is built on top of Inca foundations, attendants prepare a figure of Christ on the cross for the main procession. They crown him with a garland of *ñucchu* (Peruvian crimson flowers) and dress him up in jewels, while small choirs, known as *chayñas* or *jilgueros*, sing out in Quechua, the language of the Inca.

At 3pm, the enormous effigy is paraded through the streets, just as the mummies of Inca rulers were centuries ago. Legend has it that in 1650, when the sacred relic was first borne through the city, it brought to an end a devastating earthquake, and ever since then the figure

El Señor, the Black Christ figure garlanded with ñucchu, being paraded through Cusco on Holy Monday

has been known as El Señor de los Temblores (The Lord of the Earthquakes). *Ñucchu*, once an offering to the Inca gods but now symbolizing the blood of a Christian divinity, rain down from balconies and firecrackers herald the Christ figure's approach. At 7pm, El Señor retreats back inside the cathedral, closing the door on another year. Whether you're a believer or not, you'll be swept away by these powerful Andean Catholic celebrations.

Still going to Semana Santa, Seville?
Pick up a leaflet specifying the departure times and routes of the processions, and then stake out your vantage point early.

Getting There Alejandro Velasco Astete International Airport is 5 km (3 miles) from the centre of Cusco.

www.peru.travel

Jubilant dancing, vibrant tunes, flamboyant floats – and we're not talking about New Orleans's Mardi Gras. Embrace the beats of Mombasa Carnival, a celebration of Kenyan life.

the alternative to
Mardi Gras, New Orleans, USA

MOMBASA CARNIVAL

Kenya

Throughout November, Kenya's second largest city vibrates with the sounds of Mombasa Carnival, the biggest festival in the country. The city's streets fill with brightly costumed dancers swaying to the traditional rhythms of African drums, zithers and tambourines, and the energetic beat of Afropop.

The highlights are two parades that converge on Moi Avenue, featuring extravagant floats and hundreds of drummers. Men and women march and dance in striped kikoi and bright kanga dress. Stalls sell a plethora of street food: *vizai karai* (potato nuggets) and *mitai* (small coconut dumplings), washed down with a special brew – *mnazi* (made from coconut sap). *Kuwakaribisha Kenya* – welcome to Kenya.

Still partying at Mardi Gras, New Orleans?
Pop into Mardi Gras World, a facility that will tell you all about how the floats are designed and made.

Getting There Taxis from Mombasa's Moi International Airport take 20–30 minutes to reach the city centre.

www.tourism.go.ke

Swap the queues, high prices and frenzy of the world's biggest red carpet, the Cannes Film Festival, for the arty movies and relaxed vibes of San Sebastián, Spain's premier film festival.

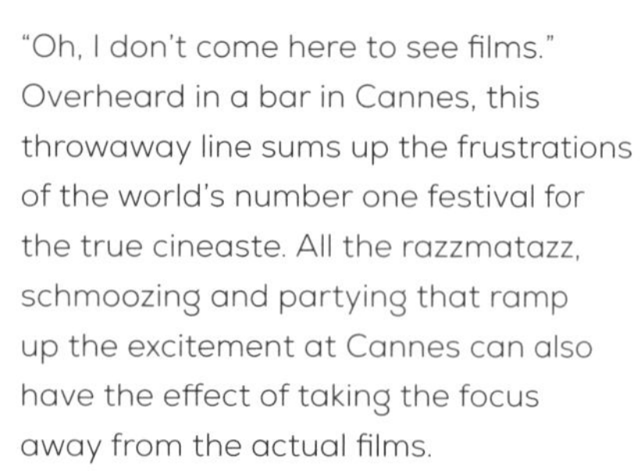

the alternative to *Cannes Film Festival, France*

SAN SEBASTIÁN FILM FESTIVAL

Spain

"Oh, I don't come here to see films." Overheard in a bar in Cannes, this throwaway line sums up the frustrations of the world's number one festival for the true cineaste. All the razzmatazz, schmoozing and partying that ramp up the excitement at Cannes can also have the effect of taking the focus away from the actual films.

San Sebastián, on the other hand, is very much a movie buff's festival. It may not have the number of weighty world premieres offered by its French cousin, but it has a good eye for new directors from all around the world. Retrospectives range from Japanese noir to individual directors to films dealing with migration.

Crucially, there's time to see such classics, as this is a more laid-back event than manic, sleep-deprived Cannes. It's also more democratic: in San Sebastián, with a little forward planning, you can buy tickets for any movie, whereas in Cannes, only films in the parallel Quinzaine section and Cinéma de la Plage beach screenings are accessible to the general public.

The Spanish festival, which runs for nine days at the end of September, is not without its glamorous side, however. Talent attending the 2021 event included Jessica Chastain, Kristen Stewart, Stanley Tucci and Marion Cotillard, and a similar roster of stars attend most years.

As well as the film festival, San Sebastián (Donostia in the local

Actor Marion Cotillard receiving an award at the opening gala of the 69th San Sebastián Film Festival

San Sebastián, with its sweeping golden beaches backed by verdant hills, at sunset

Euskara language) warrants exploration itself (this part of Spain is much less touristed than frenetic Barcelona and Madrid, after all). The festival's relaxed schedule means you'll have plenty of time time to enjoy this pretty seaside resort with its vibrant eating and drinking scene. The walk from the Kursaal (the skewed cube that is the festival hub) to the belle époque Teatro Victoria Eugenia, where press screenings take place, should take five minutes but the gauntlet of tempting *pintxos* (Basque tapas) bars on the way can easily turn it into a leisurely hour-long gourmet ramble.

Still set on Cannes Film Festival?

Plan your visit like a military campaign. Hotel rooms need to be booked at least six months in advance and, if you're planning to see any competition films, you'll need to apply for accreditation. Old films are shown for free on Plage Macé every evening.

Getting There *The closest international and low-cost hub is Bilbao, 105 km (65 miles) west of San Sebastián.*

www.sansebastianfestival.com

MOTOVUN FILM FESTIVAL
Croatia
Of Europe's "village film festivals", this one, held in late July or early August in a Croatian hill town, is among the most fun. Indie programming and a loyal following of tent-dwelling young film fans has led to people calling it "a cross between Glastonbury and Sundance".

VENICE FILM FESTIVAL
Italy
Opening the autumn awards season at the beginning of September, Venice gives Cannes a run for its money with world premieres of Oscar contenders.

LOCARNO FILM FESTIVAL
Switzerland
This Swiss festival, held in August, is perhaps the closest to San Sebastián in its mix of great setting, small-but interesting competition line-up and enthusiastic local participation. The open-air gala screenings are justly famous.

GREAT JOURNEYS

As crowds march up the Inca Trail to Machu Picchu, another ancient route through Bolivia's neighbouring mountains remains quiet. Only the most adventurous hike this dazzling path, encountering Inca ruins, rare wildlife and jaw-dropping scenery along the way.

***the alternative to** the Inca Trail, Peru*

APOLOBAMBA TREK

Bolivia

Why not give the Inca Trail a breather? Peru's well-tramped route, which winds its way through scalloped mountains and steamy cloudforests, is always buzzing with backpackers. Few, however, know that just across the border, the snow-blanketed peaks of Bolivia's Cordillera Apolobamba mountain range lay claim to some of South America's finest, and wildest, hiking. This little-known area is a bewitching place. Sacred mountains and the remarkably unchanged world of the Andes' Indigenous people await those keen to explore this less-travelled route.

There's a reason Bolivia is dubbed the "Tibet of the Americas". The five-day (93-km/57-mile) Apolobamba trail passes through the heart of the Cordillera Apolobamba and the Apolobamba Integrated Management Natural Area, an area home to dozens of 5,000-m-(16,400-ft-) plus peaks. This high-Andean terrain is untouched by tourism – instead of the Inca Trail's crowds, trekkers here have mirrored lakes, snow-polished mountains and friendly camelids for company. This is, you may have guessed, remote country; so be prepared for four nights of camping (but after one sleep beneath the vast and bejewelled skies you'll hardly be complaining).

While this path isn't stone-laid like the Inca Trail, stretches of it have their roots in Inca times. The Apolobamba's mountain-side pathways were built to grant access to now-abandoned gold mines, worked first by the Inca and then by the Spanish – and believed by historians to have helped feed the gold-fuelled myth of El Dorado. Small Inca sites are part and parcel of this route, offering a glimpse into the ancient world of this pre-Columbian civilization.

The Apolobamba trail begins in the picturesque cobbled town of Curva,

Walking through the peaty wetlands that characterize the high mountain valleys of the Cordillera Apolobamba

Above *A member of the Kallawaya people in traditional attire* ***Left*** *Local llamas grazing in the region*

where members of the Kallawaya people reside. Tracing their roots to the pre-Columbian Tiwanaku civilization and formerly employed as healers to the Inca, the Kallawaya use their remarkable ancestral knowledge of over 800 herbs to continue their craft, travelling village to village dispensing cures to the sick. It's worth organizing a guide (route-finding can be challenging on this path) and pack animals here at Curva to ensure takings are kept within, and reinvested back into, local Indigenous communities.

Climbing out of Curva, the route delivers its first stroke of magic: the snowy face of Akamani, the Kallawaya's most sacred mountain, shimmers in the sunlight. As the days fold into one another, the path meanders across the very ceiling of the earth, persisting at soaring elevations above 4,000 m (13,125 ft). Yes, it's an unruly altitude for the lungs, but these five lofty passes mean trekkers are greeted with incredible views every single day. Visible to the south is the serrated backbone of Bolivia, the Cordillera Real mountains, while the toothy, snow-snagged backdrop of the Cordillera Apolobamba rises up in the distance.

Breaking up the majestic amphitheatres of mountains are deep valleys, where glassy lakes teeming with trout and dotted with the shocking pink of flamingos provide a welcome dash of colour.

The rural settlement of Curva, located in the rugged reaches of the Apolobamba range

As the hike progresses, you emerge into pleasant pastoral scenes and settlements subsisting at unimaginably high altitudes. Tiny, alpaca-herding communities of Indigenous Aymara people (who also trace their routes to pre-Inca times) live here. Raising their animals on rough *bofedal* marshlands and rich grasslands, they inhabit thatched cottages that are built at around 4,000 m (13,100 ft). Beyond their pastures, the vicuña (the fine-haired, wild cousin of the alpaca) grazes, conservation efforts having swollen the population to over 2,500.

Camelids aren't the only wildlife that hikers encounter. Rocky outcrops conceal skittery rabbit-like viscachas, while the luckiest of visitors may even glimpse the region's most elusive resident: the jucumari, or spectacled bear, who inhabits the forests on the mountain's lower elevations.

On day four, the trek reaches its climax at the highest of the passes, the Cumbre Sunchulli. This towering, rocky juncture grants (if you're lucky with the weather) panoramas of a glaciated valley enveloped by jagged mountains and overhung with cerulean skies. A moment here makes it abundantly clear why Andes residents believe high elevations give you the closest proximity to the gods. Above you, Andean condors soar, their wide wingspans and white Tudor ruffs making them easy to spot against the blue.

After five days of climbing to the roof of the world, the trail comes to its close at Pelechuco. This huddle of tin-roofed dwellings dates back to the Spanish conquest, when it was founded as a gold-mining outpost; it now provides respite for weary hikers. Here, there's time to pause and breathe in the deep mountain air, before the bus to La Paz (Bolivia's capital) hurries your return to the city, and brings you right back down to earth.

Still want to walk the Inca Trail?

The Peruvian government limits trekkers along the Inca Trail to 500 per day, so whenever you visit, you're guaranteed to encounter the same number of people on the path. Permits sell fastest during the high season (June–August), but are easier to acquire in May – a month that's still dry and warm with high chances of catching the famous panorama of Machu Picchu.

Getting There *The nearest international airport is in La Paz. From here you can either catch a bus (which can take around 6–10 hours) on to Curva or hire a private driver.*

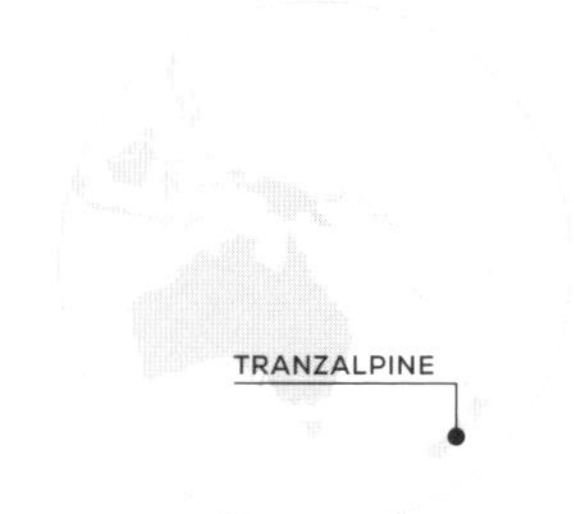

Want to whizz through epic alpine landscapes? Swerve Switzerland's popular Glacier Express *and get a front-row seat to New Zealand's thrilling mountain scenery aboard the* TranzAlpine.

the alternative to
the Glacier Express, *Switzerland*

TRANZALPINE

New Zealand

With its winding alpine passes and curling coastal roads, New Zealand is often deemed a road-tripping paradise, but this little country isn't just about the open road. Traversing the peaks of the South Island is the *TranzAlpine* train route – the most scenic (and most relaxing) way to see New Zealand's varied landscapes.

While Switzerland's *Glacier Express* navigates the serrated peaks of the Alps, New Zealand's less well-known rail adventure whisks travellers through an even greater variety of scenery. Starting in Christchurch, the train meanders through the mosaic-like Canterbury Plains before ascending into the dazzling Southern Alps (the craggy peaks that graced the screen in the *Lord of the Rings* trilogy). In the comfort of their cosy carriage, travellers can watch the scenery shift from glacier-scoured valleys, punctuated by the bright blue Waimakariri River, to dense beech forests. For a better view, you can even head to the open-air viewing car.

When you're admiring dreamy alpine views, five hours can pass in a flash. Suddenly, the train descends down the Otira Tunnel and halts at Greymouth. But instead of stepping off the *Glacier Express* into towns run with Swiss efficiency, you're in New Zealand. This means local beers, meat pies and hokey pokey ice cream, all served with down-home Kiwi charm.

The TranzAlpine *train travelling through the Southern Alps, on South Island*

Still riding the Glacier Express?

Not fussed about luxurious carriages and panoramic windows? Cheaper and less crowded trains travel the same route as the *Glacier Express*; plot your route on *www.sbb.ch*.

Getting There *You can board the* TranzAlpine *in Christchurch (on the east coast) or Greymouth (on the west coast).*

www.greatjourneysof nz.co.nz

Wild, often-deserted and complete with dramatic Scandi scenery, Norway's Helgeland Coast offers an alternative island-hopping adventure to Japan's classic cycling route, the Shimanami Kaido.

***the alternative to** Shimanami Kaido, Japan*

HELGELAND COAST

Norway

While the Shimanami Kaido, which passes over several islands in the Seto Inland Sea, can be a bit of a tourist conveyor belt, the long ride up the jagged Norwegian coast is a far less trafficked route. Norway's FV17 road winds its leisurely way between waterside towns and villages, swooping below fjords and imposing mountains as it goes. Countless islands and islets, linked by ferries or bridges, pepper the bays as if scattered from a grinder. Here, the civil engineering is as majestic as the scenery.

The Helgeland Coast section, halfway between Trondheim and Bodø, is the showpiece. The Seven Sisters peaks that loom over Sandnessjøen here threaten to split the sky with their dramatic sawtooth profile, while Torghatten – a short side-hike off the FV17 – is a mountain with a bizarre natural hole through it.

Nature may take up 99.9% of everything here, but the inhabited remaining 0.1% offers snug local places to dine and stay. Cyclists on a budget can wild camp and self-cater, enjoying this priceless landscape almost for free. Like Japan, this is a highly developed country, but less packaged: if the Shimanami Kaido is a shrink-wrapped toffee apple, the Helgeland Coast is a market-stall fresh peach.

Still cycling the Shimanami Kaido? Aim to visit between March and May in order to avoid the area's summer rains and oven-like temperatures.

***Getting There** Take a combination of train and bus from Oslo to Brønnøysund, where you organize one-way bike hire.*

www.visithelgeland.com

Mountain rocks, sandy beaches and azure waters along the Helgeland Coast cycle route

Think the fabled US Route 66 is the ultimate road trip? Think again: sun, sand and endless stretches of coastal beauty make Mexico's Coastal Route 15 the perfect choice for adventurers.

***the alternative to** Route 66, USA*

COASTAL ROUTE 15

Mexico

There are few places in the world where you can drive for hours on end with the open expanse of the ocean on one side and everything from desert landscapes to tropical forests on the other. Of the handful of such routes that still remain, the little-known coastal portion of Mexico's Route 15 is one of the best. It's a journey you can make in as little as five days or one you can savour for as long as six months. Along the way, you'll pass tiny seaside pueblos (small towns), welcoming villages, endless fields of blue agave, prehistoric landscapes, Toltec ruins and, further south, several of the world's most famous ocean resorts.

The first leg of the journey – if you choose to head south along Route 15, the approach favoured by most – begins in the inland border city of Nogales. From here, the road sweeps towards the Sea of Cortez, taking in the wild Sonora Desert scenery before cruising through the buzzing city of Hermosillo and skirting the brilliant blue sea at Guaymas. Making it this far is an adventure in itself: you'll cross paths with venomous Gila monsters, watch roadrunners whizz past and spot the occasional vulture circling overhead. This is Mexico's Wild West of legend, brought to life in front of your very eyes.

From here onwards, it's mostly a coastal road, with the shimmering Pacific Ocean as your trusty companion. Your next city stop is Mazatlán – a hundred or so kilometres (around 62 miles) on from Guaymas. Also known as the "Pearl of the Pacific", Mazatlán features more than 20 km (12 miles) of uninterrupted beaches – one of the longest stretches of sand in the world. Quaint little fishing villages and plenty of classic Mexican scenery lie ahead before Route 15 veers inward again to the east and heads toward the city of Tepic. If you don't fancy this interior stretch of the journey, continue south along the coast on Route 200 until you reach the famous southern resorts collectively known as the Big Five: Puerto Vallarta, Manzanillo, Ixtapa, Acapulco and Puerto Escondido. Live it up in these bustling areas – unwinding

Coastal scenery at San Carlos Nuevo Guaymas, a dramatic bay area on the first leg of the Coastal Route 15

on the beautiful beaches and enjoying the vibrant nightlife – before finally waving goodbye to Mexico and finishing your epic road trip at the border with Guatemala.

Still taking a trip along Route 66?

If the freewheeling stretch of Americana is still calling your name, be prepared: it's no longer a complete road. Nowadays, this legendary route is a mishmash of chopped up lanes and state roads. For advice on how to follow its original path check out the "The Mother Road: Historic Route 66" website *(www.historic66.com)*, which features step-by-step directions for retracing the former asphalt glory.

Getting There *You can either start this road trip in Nogales and head south, or you can pick up Route 200 from the border with Guatemala and head north along the Pacific Coast, joining Route 15 at Tepic.*

www.visitmexico.com

Top *Spire of the Church of our Lady of Guadalupe, the oldest church in Puerto Vallarta* ***Above*** *Locals playing dominoes on a beach in Mazatlán at sunset*

An incomparable Himalayan journey leads trekkers to the base camp of the world's most demanding mountain – and it's not Everest. Hike the fearsome Annapurna mountains for an awe-inspiring adventure.

the alternative to *Everest Base Camp, Nepal*

ANNAPURNA

Nepal

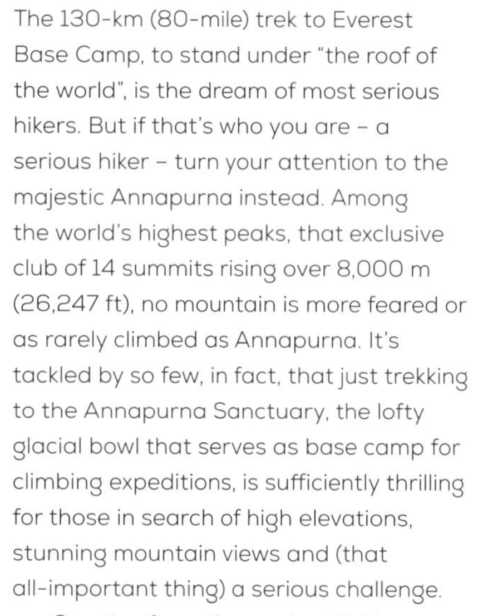

The 130-km (80-mile) trek to Everest Base Camp, to stand under "the roof of the world", is the dream of most serious hikers. But if that's who you are – a serious hiker – turn your attention to the majestic Annapurna instead. Among the world's highest peaks, that exclusive club of 14 summits rising over 8,000 m (26,247 ft), no mountain is more feared or as rarely climbed as Annapurna. It's tackled by so few, in fact, that just trekking to the Annapurna Sanctuary, the lofty glacial bowl that serves as base camp for climbing expeditions, is sufficiently thrilling for those in search of high elevations, stunning mountain views and (that all-important thing) a serious challenge.

Starting from the verdant Pokhara Valley, the trail to base camp climbs steeply through forests of rhododendron laced with orchids, bamboo and terraced rice fields, and into alpine meadows. Along the way, you'll pass several villages, traverse canyons and cross streams on rickety suspension footbridges. Leaving woodlands behind and entering a rugged realm of ice and snow, a night at the first camp gives a taste of what's to come as you rest beneath the shimmering "fish-tail" summit of Machapuchare, one of the most dramatically beautiful of all Himalayan summits.

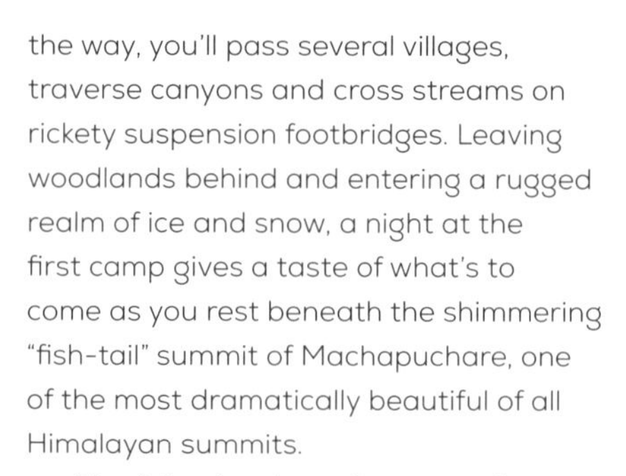

The following day, after ascending a narrow pass, you'll arrive at the Annapurna Sanctuary, a mountain amphitheatre at a height of 4,100 m (13,450 ft) and ringed by eight peaks over 7,000 m (22,965 ft), including the mighty Annapurna I, the world's most deadly mountain (so-named because of its mortality rate among climbers of around 40 per cent). Here the surrounding mountains rise up so sharply and so high that the Sanctuary receives only seven

hours of sunlight a day, even in midsummer. It's sacred to the local villagers, who consider it to be the home of the gods. After an arduous journey and with such spectacular scenery all around, you might find yourself thinking the same.

Still trekking to Everest Base Camp? Avoid the crowds by going between March and May. Instead of starting at Lukla (as most do), take the old access route from Jiri. Fewer people choose this longer trek, and it's rich with pretty villages and ancient monasteries.

Getting There *From Kathmandu, take a short flight to Pokhara. Treks to Annapurna Base Camp start from several villages close to this lakeside city.*

www.ntb.gov.np

Clockwise from top left
Trekking through Modi Khola Valley towards Annapurna Sanctuary; sherpas resting at the Sanctuary, surrounded by the mountain peaks; snowy Annapurna Base Camp in winter

Those who climb Mount Stanley – Africa's third-highest peak – are promised a breathtaking variety of flora, fauna and terrain that Africa's more famous giant, Mount Kilimanjaro, can't compete with.

the alternative to *Mount Kilimanjaro, Tanzania*

MOUNT STANLEY

Uganda

Clouds moving over the ragged peaks of Mount Stanley in the Rwenzori Mountains

To say you've climbed the fabled and challenging Mount Kilimanjaro – Africa's highest, and the world's tallest free-standing, mountain – is certainly something to boast about. But what if the journey wasn't about conquering a peak, but simply enjoying the infinite allure of Africa's natural diversity? While Mount Kilimanjaro gets all the credit, Mount Stanley offers all the majesty of its more famous rival, with more wildlife and quieter paths to boot.

Mount Stanley is the tallest of six snow-capped bergs (yep, you'll still get to experience the strange existence of near-Arctic conditions in the tropics here) that make up the Rwenzori mountain range, shared between Uganda and the Democratic Republic of Congo. Of the ten peaks on Mount Stanley, the one to tackle is the Margherita Peak, which, at 5,109 m (16,761 ft), is the highest. Make no mistake: this hike is tough – tougher, even, than that of Mount Kilimanjaro. The terrain, sliced with deep-cut gorges, rivers and ravines, offers a daily challenge. There's also more rain and lower visibility, plus some rather dicey river crossings and knee-deep mud. All in all, it's wise to organize a guide.

The rewards, however, are spectacular. The guided journey to the summit starts at Nyakalengija, passing through the plantations and homes of the Bakonzo peoples before entering the forest canopy. You'll cross overgrown tropical grasslands on thin swampy trails, a flowing tablecloth of clouds draped over the jagged grey peaks in the distance.

One of the most incredible features of tropical mountains is the extraordinary biodiversity encountered on their slopes.

***From far left** A L'Hoest's monkey; a giant lobelia standing tall; approaching the snow-capped summit of Margherita Peak*

The sheer variety of wildlife on Mount Stanley is superb, and far superior to that of Mount Kilimanjaro, where sightings are rare. Along your journey through elephant grass, some 70 species of mammal, from the forest elephant to L'Hoest's monkeys, peek out at you. Once you reach the first camp of Nyabitabe, the sound of cicadas move you to a quick sleep – a delightful way to end your first day.

As the trek continues over the next few days, the landscape unfolds to reveal a wealth of ecosystems. From the grasslands, the trail almost disappears within the dense forest, where overgrown leaves and moss-covered trees flap in your face incessantly. Yet glimpses of the mountain spur you to keep moving, as do the sounds of the Mubuku and Bujuku rivers, beckoning you to take a refreshing cold bath.

All of a sudden, the thick forest landscape opens to wide, flat plains. The mountain's abundant rainfall gives rise to weird and wonderful plants, including the giant lobelias you'll see here, which look more like miniature trees than ground cover. Wooden footbridges traverse the Bujuku River to the low Bigo bog, propelling you on to the last leg of the trek.

Around day six, the ascent to the Elena Hut begins, where one last night is spent before approaching the peak. It's a steep trek to the hut, at an elevation of 4,470 m (14,665 ft). The vegetation on the stark, ragged mountain walls slowly changes the higher up you go, revealing afro-alpine vegetation beneath your feet.

The final ascent to Margherita Peak is, as expected, tough and unyielding. Donning extra layers, gloves and head socks, and armed with an axe and crampon, you'll traverse the gnarly Elena Glacier before

making the final push towards the summit, fighting against the wily snow plains and strong winds. It's worth it: the white rocky peaks here feel out of this world. At the top, where the overwhelming views are so unspoilt and full of life, it's almost impossible to imagine descending back down to earth.

Still set on Mount Kilimanjaro?

It's the tallest mountain in the world that is accessible to hikers, so undertaking this long-distance trek is bucket-list worthy. There are many routes to the summit; avoid the most popular Machame Route and find a reputable operator to follow the taxing Western Breach route instead. It'll take longer, but you're hiking Kilimanjaro, so why not challenge yourself?

Getting There *From Entebbe International Airport, head to Kasese and arrange your trek into Rwenzori National Park, which includes an ascent of Mount Stanley.*

MORE LIKE THIS

MOUNT KINABALU
Malaysia

The jungles of Borneo, where the ascent of this 4,095-m (13,435-ft) mountain begins, are one of the world's most biodiverse places. Plants don't get more extraordinary than those dotting the three-day route here.

MAUNA LOA
USA

Rising over 9,000 m (30,000 ft) from the ocean floor, Hawai'i's Mauna Loa is the largest active volcano in the world. The route may be devoid of plants and animals, but the surreal rock formations provide a fascinating backdrop to a trek.

COPA
South America

The world's highest tropical mountain range, Peru's Cordillera Blanca contains 27 summits that are higher than Mount Kilimanjaro. Copa, at 6,188 m (20,302 ft), is the most accessible peak, and can even be skied down.

The icy wonderland of Greenland is full of glittering glaciers, magnificent moutainscapes and playful marine life. You'll be just as alone as in Antarctica in this raw wilderness, and it's much easier to get to.

***the alternative to** Antarctica*

GREENLAND

Kingdom of Denmark

Magnificent glaciers, spectacular fjords, pearlescent icebergs shaped like archways and cathedral spires. This isn't Antarctica, but it's still the end of the earth: the opposite end, to be precise. More than 80 per cent ice sheet, Greenland has one of the world's sparsest populations. But unlike Antarctica, which requires a 48-hour sea voyage from the Argentinian city of Ushuaia or a bumpy, delay-prone flight, Greenland is easily reached in a reliable four-and-a-half hours by plane from Copenhagen.

Cruising is the logical way to explore this frosty realm, which in many places has few roads. Setting out from the coastal city of Sisimiut, you'll pass basking seals and minke whales somersaulting off the shoreline. At the town of Ilulissat, a tidal icefjord glitters with bergs. Further north, dock at Qilakitsoq archaeological site to explore the ancient remains of Inuit civilizations. As you near Eqip Sermia, an active glacier, you'll see hunks of falling ice create a colossal splash as they cascade into the ink-blue water. This is a wild experience you'll never forget.

Still venturing to Antarctica?

Avoid December and January (the height of Antarctica's summer) to scoop slightly lower prices. You can shave off more cash by opting for a ship with multi-berth cabins and shared facilities.

***Getting There** Flights from Copenhagen reach Kangerlussuaq in Greenland. From here it's a short flight to Sisimiut.*

www.visitgreenland.com

Cruising past the massive Eqip Sermia glacier, one of the most active glaciers in Greenland

With towering clifftops, endless deserted beaches and fascinating ancient monuments, Ireland's Wild Atlantic Way surpasses Australia's Great Ocean Road. If you're looking for wild landscapes steeped in history, this is the road trip for you.

***the alternative to** the Great Ocean Road, Australia*

THE WILD ATLANTIC WAY

Ireland

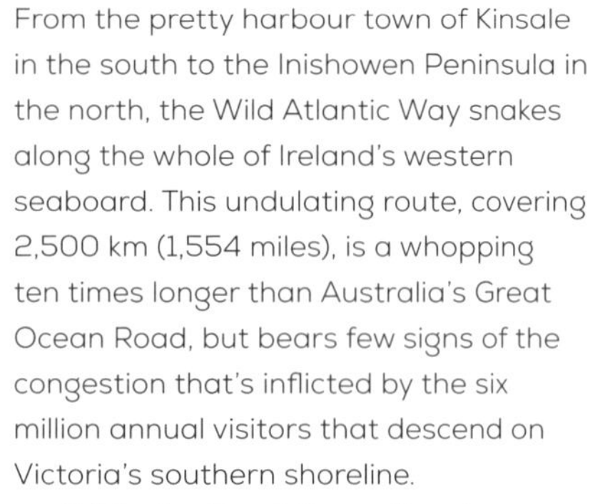

From the pretty harbour town of Kinsale in the south to the Inishowen Peninsula in the north, the Wild Atlantic Way snakes along the whole of Ireland's western seaboard. This undulating route, covering 2,500 km (1,554 miles), is a whopping ten times longer than Australia's Great Ocean Road, but bears few signs of the congestion that's inflicted by the six million annual visitors that descend on Victoria's southern shoreline.

While the Great Ocean Road was purpose-built in the 1920s, the Wild Atlantic Way developed organically over centuries. The coast-hugging road is a treasure map to some of Ireland's top natural wonders. In the south, the Iveragh Peninsula features white sandy beaches and ragged coastal outcrops. As you travel through Galway, the Twelve Bens mountain range, conical peaks that tumble down to become lush green valleys, fills your view. Moving up through County Mayo, hundreds of islands pepper Clew Bay's indigo waters. Way up north in Donegal, Glenveagh National Park promises even more natural beauty.

And there's more. Etched into the route are a millennia of historic sights. In Burren Park (County Clare), a 6,000-year-old portal tomb stands on a huge bed of limestone, while on Achill Island, in County

The 16th-century Dungaire Castle on the beautiful shores of Galway Bay

The rugged cliffs of the Iveragh Peninsula, County Kerry, a highlight along the Wild Atlantic Way

Mayo, a famine village, deserted since the 1840s, watches over the ocean from its lonely hillside setting. In Galway's Kinvara Village, a medieval castle perches on a tiny inlet on Galway Bay. These are sights worthy of stopping – this road trip might take you longer than you thought.

Still travelling the Great Ocean Road? Visitors can cut inland to take the journey in reverse for crowd control, but the best way to avoid bumper-to-bumper traffic is to hike. The Great Ocean Walk is a 110-km (68-mile) trail through glorious scenery from Apollo Bay to the Twelve Apostles.

Getting There *Cork Airport serves Kinsale on the Wild Atlantic Way's southern trailhead. Shannon Airport is midway along the route.*

www.thewildatlanticway.com

TREKKING IN BHUTAN

Trekking in northern Thailand's Chiang Mai is top of many hikers' to-do lists, but how about an adventure in the fascinating Kingdom of Bhutan instead?

***the alternative to** Chiang Mai, Thailand*

TREKKING IN BHUTAN

Bhutan

Tucked below the southern slope of the eastern Himalayas, with Tibet to the north and India to the south, Bhutan is a trekker's dream. Trails take in Himalayan peaks, high alpine slopes, seemingly endless forests and fertile valleys. Since tourism is limited (there are no armies of backpackers here) and independent travel isn't permitted, visitors will need to book a guide before their visit. But the forward planning is worth it: on your trek you'll spend time in remote villages, visit medieval fortresses *(dzongs)* and hillside monasteries, and ascend to elevations where semi-nomadic yak herders roam. Diverse, dramatic – and definitely one for the list.

Still trekking in Chiang Mai? Opt for a trek that passes through a number of villages so that you get an insight into the varied cultures of the region.

Getting There *Bhutanese national carrier Druk Air flies into Paro Airport. Be sure you have made pre-paid travel arrangements and have secured a visa in advance.*

While the Nile might be your first port of a call for a cruise, Lake Nasser has just as much to offer. This vast reservoir is surrounded by starkly beautiful scenery and some of Egypt's most astonishing monuments.

***the alternative to** cruising the Nile, Egypt*

CRUISING LAKE NASSER

Egypt

Travellers have been cruising the Nile (and writing about it – hello Agatha Christie) for centuries. However, a journey here often relies on additional coach trips to see some of Luxor's sights. What if you could skip the coach? At Lake Nasser, the sights can be viewed from your boat or they're only a short walk away from the shore.

One of the world's largest artificial lakes, Lake Nasser was created when the Aswan Dam was built in the 1960s to regulate the waters of the Nile. It's located in one of the most remote regions of Egypt, so you won't see fleets of tourist ships here; only a handful of fishers and the occasional small cruise ship or wooden felucca skim across its gleaming waters.

Honey-coloured sand dunes stretch for miles along the lake's shores, reaching Egypt's Western Desert in one direction and smothering Sudan to the south – regions inaccessible to all but the most adventurous. Nearer the lake, local Bedouins cultivate the land and tend to their animals while clusters of date palms and mud-brick houses pepper the landscape. It's unassuming scenery, until you spot the ancient Egyptian monuments.

The two magnificent temples of Abu Simbel are the highlight here. Featuring four colossal statues of the pharaoh at its entrance, the Great Temple of Abu Simbel was built to honour the pharaoh Ramses II; the smaller one to the north was created for his wife, Nefertari. These astonishing temples date back to the 13th century BC and were once carved

A small boat sailing along Lake Nasser's quiet waters en route to the temples

into the cliffs along the Nile. When the Aswan Dam was constructed, UNESCO stepped in to protect the temples from rising waters, extracting them from the cliffside and moving them in their entirety up to higher ground.

However, not all of Lake Nasser's historic treasures could be moved when the dam was created. Beneath the blue hues – startlingly clear in comparison to the silted waters of the River Nile – lie an unknown number of monuments from ancient civilizations. Lake Nasser can reach depths of around 183 m (600 ft) in some places; just imagine what treasures are submerged underwater.

Still want to cruise the Nile?

If you can't resist a trip down the Nile to the treasure trove of Luxor, book a spot on a small tour group. Fewer members means more flexibility and an adaptable itinerary that can circumvent the crowds.

Getting There *Arrive in Aswan (a city at the northern edge of Lake Nasser) via local carrier EgyptAir, or a train from Cairo or Luxor.*

www.egypt.travel

Clockwise from top left *The striking ruins of the Philae Temple of Isis, on Lake Nasser's Agilkia Island; the lake's most famous sight – the Great Temple of Abu Simbel; wandering through the Great Temple*

Dubbed "The Pride of Africa", the family-run Rovos Rail *offers the trip of a lifetime. Like the* Orient Express, *the trains capture the enchantment of the jazz age, but the wild African landscapes make the journey even more incredible.*

***the alternative to** the* Orient Express, *Europe*

ROVOS RAIL

Africa

On old railway tracks criss-crossing Africa, the vintage wood-panelled carriages of a *Rovos Rail* train chug along at no more than 96 km (60 miles) per hour. Plumes of grey steam spout from the chimney and wondrous scenery unfurls from the windows as guests sip champagne cocktails among the Agatha Christie-style furnishings. This is a nostalgic return to early 20th-century luxury train travel.

Auto-parts dealer Rohan Vos established *Rovos Rail* in 1989. Inspired by his passion for trains and the work of the Railway Preservation Society in Witbank, where Rohan ran his auto spares business, he decided to buy old train carriages and build a private family caravan that would wind around the African continent. Unfortunately, or rather, fortunately, the feasibility of his vision was tempered by the high tariffs he needed to pay to run the train so he decided to sell tickets to the public. Thus the *Rovos Rail* tourism concept was born.

Like its iconic counterpart the *Orient Express*, *Rovos Rail* exudes indulgent glamour: decadent meals, fine wines and white linen-bedecked tables in train carriages that evoke the spirit of the roaring 1920s. There's no television, radio or wi-fi on board. Instead, the ever-changing landscapes – sweeping savannahs and majestic deserts, serrated mountains and forested seascapes – provide the entertainment. On some journeys, which range from anywhere between 48 hours to 15 days long depending on the route, historians regale guests with fascinating stories of Africa's wars and fortunes.

Spotting springbok from the open balcony of the observation car

A Rovos Rail train passing through the Heidelberg grasslands in South Africa

There are plenty of journeys to choose from. One of the most stunning travels from Pretoria to Victoria Falls, taking in rugged plains and high escarpments over four days. The highlight is a stop at the golden savannahs in Zimbabwe's huge Hwange National Park. The Big Five (elephants, buffalo, lions, leopards and rhinos) all roam here. The journey ends near the world's largest curtain of falling water, known locally – and aptly – as Mosi oa Tunya, "the smoke that thunders".

Whichever route you follow, you're guaranteed to see some of Africa's most beautiful landscapes and the finest wildlife on Earth. Spotting a lion from your carriage? You won't see that from the *Orient Express*.

Still travelling on the Orient Express?
The classic London to Venice and Paris to Istanbul routes are, of course, enormously romantic but consider other options like Venice to Budapest or Vienna to London.

Getting There *Most journeys start at Cape Town, Durban or* Rovos Rail's *private station in Pretoria, South Africa.*

www.rovos.com

MORE LIKE THIS

THE CANADIAN
Canada
Made up of original 1950s stainless-steel carriages, this iconic train links Canada's east and west coasts. On its journey between Toronto and Vancouver it passes stunning prairies, lakes, mountains and towns.

THE GHAN
Australia
Running from Darwin in the north to Adelaide in the south, *The Ghan* takes you through Australia's outback landscapes and towns, all with a good dose of old-world luxury.

Looking for a scenic coastal drive? Swerve the busy Amalfi Coast and take a road trip on Corsica's wilder route, which curves along pink and burnt-orange cliffs and plunges into turquoise coves, instead.

the alternative to *the Amalfi Coast, Italy*

CORSICA'S NORTHWEST COAST

France

For panoramas so gorgeous they'll leave you speechless, drive Corsica's dizzying D81 highway, linking the island's capital, Ajaccio, with the town of Calvi. While Italy's famed yet heavily trafficked Amalfi Coast offers pretty pastel-coloured towns, this serpentine stretch (150 km/93 miles) of Corsica's northwest coast shows Europe's wilder side, squeezing between a series of astonishing rock formations and offering plenty of stunning seaside vistas. It's a wild drive too: countless zigs and zags along the route – seemingly toward the precipice and then, at the last minute, away again – will make 50 km/h feel like you're in a Formula 1 final.

From the bustling bayside city of Ajaccio, the birthplace of Napoleon Bonaparte, the D81 heads northward, gently at first. After about 50 km (31 miles) the road comes to Cargèse, an area settled in the late 1700s by Greek immigrants. Their presence explains why the village has not one but two 19th-century churches: one Roman Catholic, the other Greek Catholic. As you leave this hilly coastal town and head north, the road changes gear. The pretty hamlet of Piana marks the start of the Calanche, a 10-km (6-mile) stretch of rocky spires and vertical bluffs that rise up from the cliffside and loom over the highway – their roseate tones in stunning contrast to the brilliant blue hues of the Mediterranean beyond. You'll want to stop off here (and not only to take pictures): just off the road are a number of scenic walking trails that offer alternative angles of the rock towers.

The deep-blue waters of Porto Bay, surrounded by stunning red ochre and forested rocks

Above *Looking out onto the sea from the town of Porto*
Right *The picturesque Ponte Vecchiu, a Genoese bridge near Galéria*

The bustling harbour of Calvi, the largest city on Corsica's northwest coast and the final stop on this D81 drive

Shortly after the D81 returns to sea level, drivers arrive in the town of Porto, its harbour and pleasure port guarded by a square Genoese tower erected in the 1500s. This is a great place to spend the night, with affordable hotels and dining options, plus a range of activities. Several local boat operators here offer excursions to the stunning Scandola Nature Reserve nearby, a UNESCO World Heritage-listed treasure trove of jagged cliffs and caves that can only be viewed from the water. Alternatively, a few kilometres inland of Porto, take the one-time mule trail through the Gorges de Spelunca, which follows the Spelunca River as it swoops around boulders and fills up natural pools (ideal for a dip) on its way to the sea. Peckish after your walk? Stop for a scenic picnic at the Pont de Zaglia, an arched Genoese bridge that's been spanning the river for over two centuries.

A short drive north – complete with an exhilarating series of switchbacks – takes drivers through the village of Galéria. Highlights here warrant another stop-off: see the crystal-clear waters of the Fango River, wander through the biosphere reserve known for ancient holm oaks or spot the elegant Genoese bridge, Ponte Vecchiu.

Once you've had your fill of natural wonders, the city calls. Calvi, the largest urban centre along the northwest coast, is your final stop, but this is a leg to linger over. Here the D81 weaves across slopes smothered with aromatic maquis shrubs (whose scents waft on the sea breeze) and unleashes sweeping sea views along the way. Eventually, the road trundles into the port of Calvi, set on a crescent-shaped bay and dominated by a massive, 15th-century citadel with no fewer than five bastions. The sprawling beach here (heaps longer than any on the Amalfi Coast) is a great spot to lounge, rent a kayak, windsurf or check out snorkelling and scuba diving options. If you're a fan of scenic train trips, you can hop on the metre-gauge Chemins de Fer de la Corse railway back to Ajaccio (5 hrs). This two-car train trundles through the lowlands and hills of the Balagne region before heading into the dense forests and formidable mountains of Corsica's interior. An epic train journey after an epic road trip – what more could you ask for?

Still want to drive the Amalfi Coast? Head to Italy's most famous stretch of coastline in spring. You'll meet far fewer tour coaches along the road and the sea's just about warm enough for a dip.

Getting There *Fly into Ajaccio and hire a car here. You can return to Ajaccio after your trip or fly home from Calvi. Ferries from France are also an option.*

www.visit-corsica.com

B9 ROAD
Mauritius
Mauritius's spectacular B9 road licks the turquoise waters of the Indian Ocean as it curves around the island's south-west coast. Grand vistas and unspoilt beaches await drivers who venture to this lesser-known part of the country.

GRANDE CORNICHE
France
Clinging to the precipitous, maquis-covered hillsides of the Côte d'Azur, the Grande Corniche – which links Nice with the town of Menton – offers unsurpassed views of the dramatic meeting of land, sea and chic French seaside resorts.

PUGLIA
Italy
This laid-back jaunt around Italy's heel promises picturesque towns, including Alberobello (which looks like something out of a fairy-tale), hidden coves and a far less stressful drive than Amalfi's skinny winding route. Italian road trip: sorted.

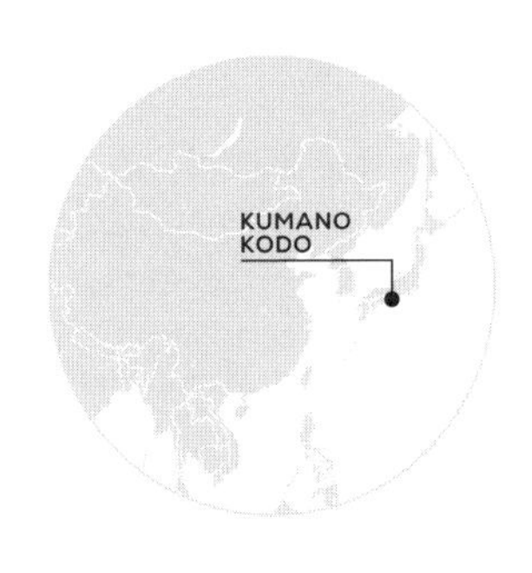

One of just two pilgrimage routes (the other being the Camino de Santiago) with UNESCO World Heritage status, the paths of the Kumano Kodo weave through the tranquil forests of Japan's spiritual heartland.

The picture-perfect view of the Kumano Nachi Taisha shrine, with the Nachi Falls cascading behind

the alternative to
Camino de Santiago, Spain

KUMANO KODO

Japan

In 2004, almost 20 years after UNESCO gave the Camino de Santiago World Heritage status, a second set of pilgrimage trails got the UNESCO nod. Just as old and just as rewarding, the Kumano Kodo network has been leading pilgrims to the three Kumano Sanzan grand shrines for more than 1,000 years.

The Kumano Kodo's sacred paths are sketched across Honshu's rural Kii Peninsula, an area known as the "land of gods" (in local Shinto beliefs, deities reside in every tree, river and waterfall here). A journey along one of the trails is traditionally considered a route to spiritual rebirth.

Walking the old cobbled pathways of the Kumano Kodo in traditional dress

Unlike Spain's famous route, where you'll find yourself dodging cyclists and trying to overtake sprawling walking groups, the Kumano Kodo trails offer an unhurried and largely uncrowded way to experience Japanese culture. On the most-walked route, the 68-km- (42-mile-) Nakahechi, old cobbled pathways undulate through the mountains, passing tiny villages and friendly *ryokan* (guesthouses), complete with hot-spring baths, along the way. Despite being the most popular trail, this wooded route, dappled with sunlight and surrounded by waterfalls, is still wonderfully hushed.

Other paths include the far more arduous Kohechi trail, which promises a leg-busting up-and-down trek beginning at the historic temples of Mount Koya. The temple lodgings here offer an immersive encounter with Buddhist Japan: visitors

can fill up on the vegan *shojin-ryori* meals and join the monks for their morning rituals.

Whichever of the half dozen routes you opt for, all end at the Kumano Sanzan grand shrines. While each shrine dazzles in its own right, the view at the Kumano Nachi Taisha shrine, of a vermillion pagoda set against the tumbling Nachi Falls, is the jewel in Kumano Kodo's crown, and it's worth every step of the journey.

Still walking the Camino de Santiago?
If you want to trek to the Cathedral of Santiago de Compostela on a less-trodden path, try the Coastal Camino Portugués route along the Atlantic coast.

Getting There *Fly to Nanki-Shirahama Airport, from where bus and train networks connect to trails.*

www.tb-kumano.jp

KINTYRE 66

A compact version of the NC500, the Kintyre 66 has the dramatic coastal roads and similarly striking scenery, but with far less traffic.

the alternative to *the North Coast 500, UK*

KINTYRE 66

UK

Looking for an alternative Scottish road trip? Sure, the country's famous North Coast 500 is packed with gorgeous scenery, but you'll need nearly a week to do it justice, whereas the Kintyre 66 could be tackled in half a day. But what's the rush? This 106-km (66-mile) road trip around Scotland's most southwesterly peninsula is all about slowing down and embracing local life.

Bustling harbours and isolated fishing villages serving up super-fresh seafood make charming stop-offs here. As do the numerous whisky distilleries and wave-lashed beaches. Caves at Keil and castles in Saddell and Skipness are also striking additions to the glorious scenery – though few views are finer than the setting sun over the nearby islands of Islay and Jura.

Still doing the NC500?
Starting and finishing in Inverness, the North Coast 500 gets especially busy in the summer, so aim to drive by in the spring or autumn.

Getting There *Around a three-hour drive from Glasgow, the Kintyre 66 forms a loop from Kenna-craig to Campbeltown – both have ferry terminals.*

They may not be household names the world over like their Greek counterparts, but Montenegro's islands also offer sandy shorelines, as well as spectacular nature, myth and history. And all with a fraction of the crowds and at a fraction of the cost.

***the alternative to** the Greek Islands, Greece*

MONTENEGRO'S ISLANDS

Montenegro

Not got the time for an epic Greek odyssey? Try Montenegro's dozen or so islands instead, which are mercifully close together and no less breathtaking. Most lie in the warm embrace of the compact Bay of Kotor, while others line the edge of the Adriatic, just a short boat ride (or even walk) from the shore. From lushly forested islets fringed by gorgeous sweeps of sand to a human-made stack of rock topped with a 17th-century church, each Montenegrin island offers something different.

At the southern tip of Dalmatia, a little southeast of Dubrovnik in Croatia, the Bay of Kotor is one of Montenegro's biggest draws. And it's little wonder. Made up of inlets linked by narrow passages, the bay is fringed by quaint villages, overlooked by the gloriously green Dinaric Alps. Out in the bay itself lie some of the country's most beautiful islands. While most of these are natural, the title of most iconic goes to an artificial island, the Gospa od Škrpjela (Our Lady of the Rocks). Legend has it that a group of seamen found an icon of the Madonna and Child on a rock in the sea in 1452, so set about creating an island on which they could build a church to house it. They added new rocks to the pile every day, along with the odd scuttled ship, and after (give or take) 180 years, they had a fully formed, church-ready island.

Fast forward around 400 years, and the Church of Our Lady of the Rocks is one of Montenegro's best-known attractions. Small boats ferry passengers from Perast on the mainland to the island harbour in five to ten minutes, from where it's just a few steps to the pretty stone church. Inside, see stunning, 17th-century Baroque paintings and a gorgeous tapestry embroidered with a mix of golden and silver fibres – as well as, so they say, strands of the artist's hair.

The beautiful Bay of Kotor, surrounded by forested mountains and the medieval town of Kotor

Left *The human-made islet of Our Lady of the Rocks with its old church* ***Above*** *The entrance to the small church of Our Lady of the Rocks*

***Above** Kitesurfing in the waters off Ada Bojana **Right** The public beach on Sveti Stefan, perfect for swimming*

Our Lady of the Rocks is far from the only church-on-an-island here. You can also explore Sveti Đorđe (Saint George), a natural island home to a picture-perfect 12th-century monastery and a graveyard. There's also Ostrvo Cvijeća (Island of Flowers), connected to the mainland by a narrow strip of land, and featuring a 13th-century abbey. And let's not forget the tiny Gospa od Milosti (Our Lady of Grace), home to a spectacular 15th-century Franciscan monastery.

But it's not all about houses of God. Sveti Marko, the biggest island, doesn't have any churches – or, in fact, anything human-made. This is an island entirely covered in greenery, rimmed by beautiful beaches. It's off most tourist radars, but boat owners in Tivat may be persuaded to ferry you there for a small fee.

As well as the Bay of Kotor, Montenegro has a host of islands dotted along its Adriatic coast. Two are so close to the mainland that you don't even need to charter a boat. Sveti Stefan, a postcard-worthy islet, is connected to the mainland by a narrow tombolo. While most of the island is occupied by a resort, the beach on the left side is free and open to the public, its turquoise waters perfect for swimming and snorkelling. Ada Bojana is situated in the far south of the country. Created by a river delta of the Bojana River, this island is connected to the mainland by road bridge. With its long stretch of idyllic beach and subtropical climate, it's a great place for swimmers, windsurfers and kitesurfers.

Looking to venture further out to sea? Head for Sveti Nikola Island, Montenegro's largest island in the Adriatic, which lies a couple of kilometres (a mile) out from the mainland town of Budva. Breaching the surface of the sea like a shark's fin, the island has soaring rocky cliffs – but only on one side. At the northern tip of the lopsided isle is a sandy beach, while its densely forested interior is roamed by fallow deer. Hire a kayak to explore the coves, including a secluded beach at the southern end.

With a mere 88 km (55 miles) between Our Lady of the Rocks in the west and Ada Bojana in the east, you can see all of Montegnegro's islands in a long, leisurely weekend. How's that for time-poor island-hoppers?

Still hopping around the Greek Islands? Instead of Crete, Corfu and Rhodes, opt for lesser-visited but just as lovely islands like foodie haven Sifnos, proudly traditional Tinos and away-from-it-all Amorgos.

Getting There *Fly into Podgorica in Montenegro or Croatia's Dubrovnik and catch a bus to Budva, a good base for visiting islands in all directions.*

www.visit-montenegro.com

STOCKHOLM ARCHIPELAGO
Sweden
Outdoor enthusiasts may find it hard to choose which of the 20,000 islands and islets east of the Swedish capital to hop around. Our favourites are Grinda for swimming, Svartsö for biking and Sandön for kayaking.

FIJI ISLANDS
Fiji
Made up of hundreds of islands spread over thousands of square kilometres, Fiji is the ultimate island escape. Whichever islands you sail to, you're guaranteed lush landscapes, golden shores and loads of stunning sea life.

OUTER HEBRIDES
UK
From the Neolithic stone circles and picture-perfect beaches of Lewis and Harris to the skyscraping cliffs and swooping sea birds of St Kilda, Scotland's western isles are utterly unique. Hop between them on the regular ferry services.

ARCHITECTURAL MARVELS

With its brilliant cupola mosaic and crypt, the principal cathedral of the Serbian Orthodox Church, Saint Sava in Belgrade, rivals the Hagia Sophia in Istanbul as a masterpiece of religious architecture.

the alternative to *the Hagia Sophia, Turkey*

CHURCH OF SAINT SAVA

Serbia

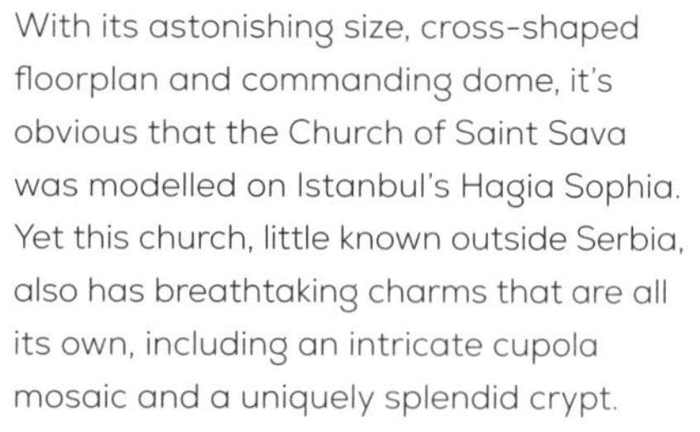

With its astonishing size, cross-shaped floorplan and commanding dome, it's obvious that the Church of Saint Sava was modelled on Istanbul's Hagia Sophia. Yet this church, little known outside Serbia, also has breathtaking charms that are all its own, including an intricate cupola mosaic and a uniquely splendid crypt.

It may only be the second-largest church built for the Orthodox Christian faith (after you-know-where in Turkey), but few other churches dominate their city like Saint Sava towers above Belgrade. Whether you're promenading along the waterfront, climbing up Belgrade Fortress or shopping in Kalenić Market, you'll always find the church's mammoth dome peeking out over the city's rooftops – and providing valuable directional assistance for locals and visitors alike.

Venture a little closer and the Church of Saint Sava only becomes more impressive. With its foundations laid in 1935, on the supposed site where the remains of the namesake Serbian monk were burned by the Ottoman occupiers in 1594, the cathedral looms over the Vračar plateau in the heart of the city. Admire the elegant façade from every angle with a stroll around its four rigorously symmetrical sides, and peer up at the enormous main dome with its giant gold-plated cross.

Of course, the building's full splendour can only be appreciated after you step inside. The vast interior, illuminated by sunlight streaming through the dome above, can accommodate a frankly absurd 10,000 worshippers. You'll immediately find your eye drawn upwards to the cupola, the inner vault of the dome, as it's home to a magnificent mosaic depicting the Ascension of Christ. A masterpiece of Orthodox art, its millions of tiles, all curved around a length of 1,250 sq m (13,455 sq ft), also make it a great feat of engineering.

The graceful marble façade of the Church of Saint Sava, built in Serbian-Byzantine style

Top The golden interior of the crypt, adorned with mosaics and frescoes *Above* The splendid Ascension of Christ mosaic in the shimmering dome

It's tempting to keep looking up forever, but an even greater treasure lies beneath your feet. Descend the marble staircase into a spectacular subterranean world quite unlike any other part of the church – or, in fact, of any church. With its shimmering gold ceiling, supersized chandeliers, Murano glass mosaics, and bright frescoes, this may be the most gloriously decorative crypt in Christendom.

For all these remarkable features, the church is still not quite complete – as evidenced by the scaffolding inside. It should be finished by 2030, but you won't want to wait until then. Even as a work in progress, Belgrade's Church of Saint Sava is an absolute must-see.

Still going to the Hagia Sophia?
You'll need time to explore it all, so come for the 9am opening. Start with the upstairs galleries and you might even get a corner (briefly) to yourself.

Getting There *A taxi journey from Belgrade Nikola Tesla Airport to the city centre takes about 30 minutes.*
www.hramsvetogsave.com

Like Gaudí's Sagrada Família, the Cathedral of Rio de Janeiro is ground-breakingly unconventional. But unlike Barcelona's unfinished masterpiece, you can wander inside it for free.

***the alternative to** the Sagrada Família, Spain*

CATHEDRAL OF RIO DE JANEIRO

Brazil

Rising from the centre of Rio de Janeiro is a huge pyramid. But this is no ancient Mayan structure – it's the city's futuristic cathedral, an epic take on a religious structure as astonishing as Barcelona's Sagrada Família. Designed by Edgar Fonseca in the 1960s, Rio's massive cathedral is a modernist masterpiece

Stepping into this cool sanctuary, you are submerged in a wash of vivid reds, blues, yellows and greens, emanating from four panels of stained glass that stretch 64 m (210 ft) high. There are no obvious corners; instead it's all bending lines, from the circular floorplan to the curved altar steps. Much like Gaudí's Spanish vision, this astonishing building is not only a must for the devout, but also for all fans of extraordinary architecture.

The Mayan pyramid-like Cathedral of Rio de Janeiro

Still going to the Sagrada Família?

Book a place on a guided tour for faster entry and fascinating insights. And visit at sunset, when the golden light hits the stained glass, creating a divine-like glow.

***Getting There** Shuttle buses run every 30 minutes from Rio de Janeiro International Airport to the city centre, taking about one hour.*

www.catedral.com.br

MORE LIKE THIS

Hallgrímskirkja
Iceland
Reykjavík's tall cathedral resembles a giant space rocket flanked by volcanic basalt. Started in 1945, the construction took 41 years to finish.

CARDBOARD CATHEDRAL
New Zealand
Yes, Christchurch's Transitional Cathedral is actually made of cardboard tubes (and some wood, steel and glass). The A-frame's humble design gives it an unusual beauty.

Both resembling beautiful blooms, and both Bahá'í houses of worship, Santiago's Bahá'í Temple and Delhi's Lotus Temple have won numerous architectural awards.

***the alternative to** the Lotus Temple, India*

BAHÁ'Í TEMPLE

Chile

If you were a large herbivorous dinosaur, you would undoubtedly find Santiago's Bahá'í Temple irresistible. Elegant torqued wings curl upwards like the petals of some colossal flower, which looks all the more alluring against the backdrop of the foothills of the Andean Mountains. For humans, the 30-m- (98-ft-) high temple resembles less of a culinary temptation and more of a huge sculptural work of art.

Like Delhi's Lotus Temple and all other Bahá'í houses of worship around the world, the shrine in Santiago has a nine-sided shape. The nine arching "petals" are ringed by nine pathways with nine reflecting pools that accentuate the exterior's dramatic cast-glass panels. The interior, sheathed in filigree marble, echoes the intricate weave of Japanese baskets. Curved forms, inspired by the shapes made by Sufi dancers, swirl together, with the light inside shifting from silver to ochre to violet as the sun sinks towards evening.

The temple, so deserving of its accolades, exudes a sort of primordial beauty, undoubtedly captivating all who set eyes on it.

Still visiting the Lotus Temple?

A great base nearby is the Eros Hotel, which has views of the Lotus Temple from the upper floors.

***Getting There** The Bahá'í Temple is an hour's taxi ride east from the Santiago de Chile International Airport.*

www. bahai.cl/ templobahai

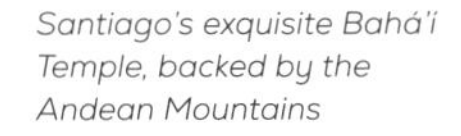

Santiago's exquisite Bahá'í Temple, backed by the Andean Mountains

Clockwise from top *The elaborate main entrance to Notre-Dame d'Amiens, featuring impressive carvings on the portals; the cathedral's elegant Gothic nave, lined with statues; the towering exterior, seen from across Place Notre-Dame*

You'll find France's most famous Notre-Dame in Paris. But France's most impressive cathedral? That'd be another Notre-Dame, this time in Amiens. This hulking masterpiece of Gothic architecture is bigger and better than its Parisian counterpart.

the alternative to *Notre-Dame de Paris, France*

NOTRE-DAME D'AMIENS

France

You're probably familiar with the soaring towers of Paris's iconic Notre-Dame. Now, imagine a building so cavernous it fits two of the capital's cathedrals within its hallowed halls. Step up, Notre-Dame d'Amiens. France's largest and tallest cathedral dominates the skyline of Amiens, and has been hailed a Gothic masterpiece by UNESCO, who awarded it World Heritage status in 1981.

This massive cathedral was built in just 50 years (between 1220 and 1270) – but scrimping on time didn't mean scrimping on decor. The exterior of Notre-Dame d'Amiens is a feast for the eyes. Here, embellishment has gone into over-drive, with three portals busy with biblical figures (including saints, magi, apostles and angels), and above them, 22 stony, life-size kings. Inside is a 42-m- (138-ft-) high ceiling, an enormous nave, four aisles and a number of radiating chapels. This vast interior may be a little barer than the cathedral's overblown façade but its sheer size makes it intensely luminous. Light floods through the nave, spotlighting the array of statuary inside and making the Notre-Dame de Paris look, in comparison, rather sombre.

Still want to see Notre-Dame de Paris?
As a result of the devastating 2019 fire, Paris's cathedral is closed, with restor-ation work predicted to continue until 2024. To glimpse the cathedral's exterior you can take a river tour along the Seine or head to Place Jean-Paul II for up-close views.

Getting There *Amiens is around two hours by car from Paris via the A16 and A1. Regular trains (1 hr 30 mins) also run from Paris to Amiens.*

www.visit-amiens.com

Uncannily similar in style and shade of colour, Lisbon's Ponte de 25 Abril more than matches the Golden Gate Bridge's beauty, and its visitor centre adds some brains to boot.

the alternative to the Golden Gate Bridge, USA

PONTE 25 DE ABRIL

Portugal

Named after the Carnation Revolution, the day in 1974 when a military coup overthrew Portugal's Estado Novo authoritarian regime, Lisbon's Ponte 25 de Abril is the longest suspension bridge in Europe. As it's painted in the same orange vermillion as San Francisco's Golden Gate Bridge, it looks remarkably like its more famous (and much more visited) predecessor. In fact, you can easily mistake it for the Californian icon in photos.

And the similarities don't end there. In the same way that the Golden Gate ferries people out of San Francisco and into the relative wilds of Marin County, the Ponte de 25 Abril connects Lisbon to the district of Almada, on the opposite side of the River Tagus, and the long stretch of sand that runs down the Costa da Caparica beyond – though the balmy waters that lap the coast here are a far cry from the chilly surf that crashes onto the Marin Headlands.

Where the two bridges noticeably differ, however (aside from the minor drawback that you can't cross the Ponte 25 de Abril on foot), is in their visitor centres. At Pilar 7, housed right inside one of Ponte de 25 Abril's supporting pillars, you can get an up-close insight into its construction through multimedia displays, before catching a lift up to a glass-floored observation deck for spectacular views along the river. It's a far more satisfying experience than a trip to the Golden Gate Bridge Welcome Center, which is located several hundred metres from the bridge

Ponte 25 de Abril, the longest suspension bridge in Europe, spanning the Tejo Estuary

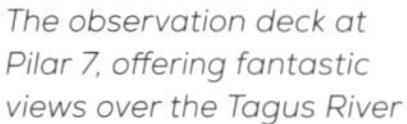

The observation deck at Pilar 7, offering fantastic views over the Tagus River

itself and is mostly given over to a gift shop stocked with tonnes of themed merchandise. Who needs yet another souvenir T-shirt when you can sneak a peek at the inner workings of an engineering marvel?

Still going to the Golden Gate Bridge?
The best way to take in the Golden Gate Bridge is from the seat of a bicycle. You can rent bikes in Fisherman's Wharf, cycle along the waterfront and over the iconic bridge, and then catch the ferry back across San Francisco Bay from the seaside city of Sausalito.

Getting There *Flights to Lisbon land at Humberto Delgado Airport, and trains arrive at the central Santa Apolónia station.*

www.visitlisboa.com

Quieter than the hugely popular Schloss Neuschwanstein, Schloss Lichtenstein – the "fairy-tale castle of Baden-Württemberg" – is no less romantic than its Bavarian counterpart.

the alternative to
Schloss Neuschwanstein, Germany

SCHLOSS LICHTENSTEIN

Germany

Although it's considerably smaller in size, it's immediately obvious why Schloss Lichtenstein draws comparisons to Bavaria's Schloss Neuschwanstein. A Neo-Gothic castle perched on a rocky outcrop? Check. Beautiful surrounding countryside? Double check.

But while Schloss Neuschwanstein is one of Germany's most-visited tourist sites, Schloss Lichtenstein has managed to avoid being engulfed by international tour groups. It's a mystery why: the grand interiors and epic façades are more than worthy of tales featuring dragons, knights and princesses. There's even an (apparently friendly) ghost named Alfons – keep an eye out as you explore.

Still going to Schloss Neuschwanstein?
Avoid the day-trippers from Munich by advance booking a tour for early in the morning or late in the afternoon.

Getting There *The nearest city is Stuttgart, from where it is possible to get a train or bus to Honau and then walk. You can also drive to Schloss Lichtenstein.*

www.schloss-lichtenstein.de

Surrounded by buildings that are almost as astonishing, Santiago Calatrava's Neo-Futurist opera house in Valencia puts even Sydney's in the shade. Catch a performance of zarzuela (a type of Spanish operetta) to experience it at its finest.

the alternative to *the Sydney Opera House, Australia*

PALAU DE LES ARTS REINA SOFÍA

Spain

One of the most recognizable buildings in the world, the Sydney Opera House is an Aussie icon. The staggered white shells of its famous roof look like billowing sails and have helped propel the building – and the city in which it sits – onto the front cover of a thousand-and-one tourism brochures. When Danish architect Jørn Utzon won New South Wales's competition to design a performing arts centre in 1957, he could not have known the legacy he would leave. The Sydney Opera House changed the image of Australia, and today 11 million people pass through its doors each year.

On the other side of the world, on Spain's southeastern coast, a similarly ground-breaking opera house has been revitalizing another city's fortunes. The sensational Palau de les Arts Reina Sofía, one of a number of Neo-Futurist designs that make up the City of Arts and Sciences complex, epitomizes Valencia's 21st-century reinvention. When Valencian-born "starchitect" Santiago Calatrava won the commission to regenerate what was then a rundown and neglected area, he set about designing buildings that were as much sculptural structures as they were functional performance spaces. This ensemble of cutting-edge architecture, which runs for 2 km (just over a mile) along the lush Jardín del Turia, has become one of the most stunning sights in Spain.

The complex is a striking prospect, rearing out of the dried-up riverbed that curves around Valencia's Old Town and dwarfing the surrounding neighbourhoods. And the dazzling Palau de les Arts Reina Sofía is the jewel in the crown. For visual impact, it certainly rivals the Sydney Opera House, its blindingly white dome jutting out into a shallow pool and resembling an enormous cracked egg or, with its curling plume, a giant warrior's helmet. ▸

The futuristic Palau de les Arts Reina Sofía, rising majestically from the former bed of the Turia River

Above The shell-shaped Auditori, the main auditorium
Right The stunning planetarium and science museum in the City of Arts and Sciences complex

In a nod to Valencia's history of ceramic production, thousands of *trencadís* (broken mosaic tiles) were used to decorate its surfaces. The effect is spellbinding, grabbing your attention completely. This is saying something in a place like the City of Arts and Sciences, where the science museum resembles the bleached bones of a whale, the planetarium looks like a huge eye and the complex's events space is shaped like an enormous Venus flytrap that has sprung firmly shut.

But this isn't a case of style over substance. As a performing arts centre, the Palau de les Arts Reina Sofía also delivers. Its emblematic Sala Principal is the venue for opera, ballet and *zarzuela*, a very Spanish kind of musical theatre that combines song with speech, much like an operetta. Although originally from Madrid, this is a national art, and you should try to get tickets for a show if you can; Francisco Asenjo Barbieri's *El Baberillo de Lavapiés* is considered the classic *zarzuela* and is a favourite of Les Arts's programme manager.

There are four venues in total, which together seat 3,000 people, around half the capacity of the Sydney Opera House. Size isn't everything, though, and you'll get a much more intimate performance at Les Arts. A series of panoramic lifts, which are actually set inside the building's walls, take visitors to the Auditori, the main auditorium, which, with its shell shape, has superb acoustics. The views widen the higher up you go, reaching out over the opera house's palm-tree terraces, across Valencia and down towards the sea. If you can't see a performance at Les Arts, guided tours are the next best thing. They pay a visit to the Centre de Perfeccionament training centre for young singers and take you inside the Sala Principal and the Auditori; on some tours, you can even sit in on a rehearsal.

They say you can't put a price on art, which is a good thing, because the Palau de les Arts Reina Sofía cost nearly €480 million to build, four times the estimated amount. But that pales in comparison to Jørn Utzon's blowout on the Sydney Opera House, which came in over 15 times its original budget. Utzon quit the job, and the country, seven years before the Opera House was completed. He never did get to see his building in person, unlike the 155 million people who have in the years since. That's 155 million and counting...

Still going to the Sydney Opera House?
Sunset is a beautiful time to visit, when the dimming light plays on the building. You'll also be able to catch the *Badu Gili* light projection of First Nations art on the Opera House's eastern Bennelong sails.

Getting There *Valencia Airport is 8 km (5 miles) from the city centre. There are frequent metro and bus links.*

www.lesarts.com/en

HARBIN OPERA HOUSE, China
This sinuous building was designed by MAD, the Beijing-based architecture firm led by Ma Yansong. With its tentacle-like side buildings, it looks like it's slipped out of the Songhua River. The wood shell interior is equally dramatic.

GUANGZHOU OPERA HOUSE China
In the shape of two giant rocks smoothed by water, Zaha Hadid's opera house is spectacular at night, when its reflection shimmers on the surface of the Pearl River.

OSLO OPERA HOUSE Norway
The angled outside of the Oslo Opera House is covered in marble and white granite, resembling a glacier rising out of a fjord. It slopes down toward's the city's harbour and was purposely designed so you can walk on its roof.

Complete with colourful mosaic tiles and an opulent gilded interior, the Tilla Kari Madrassa gleams a little brighter than St Peter's Basilica, the Vatican's most famous church. If there was ever a beauty contest for buildings, we'd bet on this madrassa any day.

the alternative to *St Peter's Basilica, Vatican City*

TILLA KARI MADRASSA

Uzbekistan

The grand Registan is to Uzbekistan as St Peter's Basilica is to the Vatican: an iconic religious and architectural monument that has come to symbolize a country. Samarkand's dramatic square, a UNESCO World Heritage Site no less, is home to three ornate madrassas (Islamic universities) – the most striking of which is the Tilla Kari Madrassa. St Peter's may be a colossal structure, its dome soaring into the sky, its façade flanked by columns, but Uzbekistan's dazzling Tilla Kari has the edge when it comes to sheer decadence.

Built in the mid-17th century by the Shaybanid General Alchin Yalantush Bahadur, the Tilla Kari is the newest, largest and most elaborate structure in the Registan. The madrassa takes its name from the mosque inside its courtyard: Tilla Kari means "gold-covered", a reference to the building's shimmering gilded dome and walls. Sure, St. Peter's dome is higher but the extravagance of the Tilla Kari is so great it'll make you dizzy. Bands of Arabic calligraphy, glazed mosaic tiles and a liberal coating of gold leaf make for a jaw-droppingly over-the-top interior. If your head wasn't already spinning, think on this: the inside of the dome is actually flat. It appears curved when you're standing under it due to an optical illusion achieved with trompe l'oeil.

And it's not just the Tilla Kari's mosque that captures photographers' eyes. The madrassa's 120-m- (394-ft-) long façade, a riot of colourful majolica tiles cut into an intricate jigsaw of geometric patterns and Qur'anic script, is also a sight to behold.

Lovingly decorated, the Tilla Kari was once the city's main place of prayer; however, when Uzbekistan became part of the Soviet Union, the madrassa and its mosque were re-classified as a historic monument. Though no longer a place of worship, the Tilla Kari is still full of life. Around the central courtyard, the vaulted

The three elaborate madrassas dominating the Registan, the main square in Samarkand

tiled cells (which once contained student classrooms and accommodation) have been meticulously restored. Each now buzzes with merchants selling brightly painted tea sets, woven silks with the distinctive zig-zag pattern of ikat and miniature paintings by some of Samarkand's master artists. This lively bazaar (which hums with chit chat and promises a constant supply of hot green tea) makes for an exciting end to your grand tour of Tilla Kari. Why hasn't this thrilling site always been on my bucket list, you ask? Good question. We'd say it's time to fix that.

Still want to see St Peter's Basilica?
To see the Vatican's enormous church at its most tranquil, aim to visit outside of Mass hours (Mass takes place hourly between 9am and 12pm, and at 5pm).

Getting There *Fly directly into Samarkand airport or take the high-speed train from Tashkent (Uzbekistan's capital has a wider range of connecting flights).*

www.uzbekistan.travel

Top *Gorgeous blue and green mosaics inside Tilla Kari Madrassa* ***Above*** *Exquisite tiling surrounding the doorways*

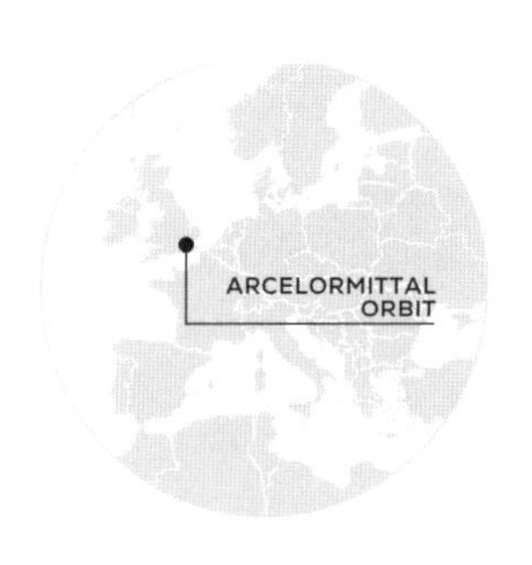

With stellar views of the capital's skyline, the ArcelorMittal Orbit does for London what the Eiffel Tower does for Paris – but with the added fun of a thrilling tunnel slide.

the alternative to *the Eiffel Tower, France*

ARCELORMITTAL ORBIT

UK

From the top of the Eiffel Tower, Europe's most romantic city stretches out before you, the silhouettes of the Louvre, Montmartre and Notre Dame poking above its skyline. The view from the summit of London's ArcelorMittal Orbit, the UK's tallest sculpture, is equally superb – the Shard, the Gherkin, Big Ben and many other icons that don't have nicknames fill a horizon that stretches 32 km (20 miles) into the distance.

The Eiffel Tower may be all about elegance, but at the ArcelorMittal Orbit the emphasis is firmly on fun. A legacy of the 2012 London Olympics, the red lattice tower designed by Anish Kapoor reflects the energy of the city, and looping around it a dozen times is the world's longest tunnel slide. It takes 40 seconds to whoosh to the bottom, alternating between views of the Queen Elizabeth Olympic Park and pitch-black sections before ending in a corkscrew. If that's not enough, you can even nip back to the top and abseil down.

Still heading up the Eiffel Tower?
Visit Tuesday–Thursday, and take the stairs (up and/or down) to avoid waits at the lift.

The dramatic red steel ArcelorMittal Orbit, with thrilling twists and turns

Getting There *The DLR from London City Airport takes about 20 minutes to reach Queen Elizabeth Olympic Park.*

www.arcelormittalorbit.com

Though smaller in size than the Empire State Building, the Louisiana State Capitol has an equally glorious Art Deco façade, a fascinating history and sweeping views over America's mightiest river.

***the alternative to** the Empire State Building, USA*

LOUISIANA STATE CAPITOL

USA

Soaring high above the riverfront city of Baton Rouge, the Louisiana State Capitol looms like a mini Empire State Building, with its Art Deco design and limestone-clad exterior. But unlike its cousin in New York City, there's never a wait to ascend to the top, and you won't have to break the bank to enjoy the views – Louisiana rolls out the welcome mat with free admission.

Sometimes referred to as "Huey Long's monument", the 34-storey building was the dream of Louisiana's controversial and visionary governor. When it opened in 1932, Long was smitten, saying, "Only one building compares with it in architecture. That's St Peter's Cathedral in Rome."

The view from the open-air observation deck on the 27th floor offers a mesmerizing sweep that takes in the Capitol Gardens, the broad Mississippi River and the vast green expanse of Cajun wetlands beyond. You might just find yourself falling in love with America's tallest state capitol, Huey Long-style, as you enjoy one of the finest panoramas in the south.

The grand Louisiana State Capitol, built in the 1930s and surrounded by lush gardens

Still craving the Empire State Building? You'll find some of the most jaw-dropping views of New York's icon from the Archer Hotel's Empire View rooms or rooftop bar.

***Getting There** It's a 70-minute drive from New Orleans International Airport to Baton Rouge.*

www.visitbatonrouge.com

Beijing's Forbidden City may have sheer size on its side, but India's smaller ancient city at Fatehpur Sikri has all the atmosphere. Purpose-built by Akbar the Great, the capital was abandoned centuries ago; now, it's the graceful ghost of a moment in India's history.

the alternative to
the Forbidden City, China

FATEHPUR SIKRI

India

Beijing's Forbidden City is one of the most popular tourist attractions in the world, squeezing in around 24,000 people a day. By contrast, Fatehpur Sikri, an Indian city frozen in time, has plenty of room for its trickle of in-the-know visitors, who can't help but linger in this eerie palace complex.

Fatehpur Sikri wasn't the Mughals' capital for very long – just 16 years compared to the Forbidden City's six centuries of imperial rule. It was commissioned by the great Mughal emperor Akbar in 1569 and, in keeping with his tolerance of multiple religions, the palace combines Hindu and Muslim elements in a harmonious fusion; traditional Indian *chhatris* (the domed gazebos on the top corners of the buildings) are effortlessly paired with Islamic pointed arches and domes. Court life played out here happily until Akbar suddenly left Fatehpur Sikri for a campaign; and a few years later it was completely abandoned due to water supply issues. To wander the site now is to see it suspended in time: red sandstone pavilions wait for the return of the royal household, while in the "Golden House", the home of Akbar's mother, faded paintings still grace the walls. Nearby, the Wind Palace's top-floor *chhatri*, designed to catch the evening breeze, hasn't lost any of its air of romance.

Unlike the pavilions of the Forbidden City, those of Akbar's dream capital vary constantly in design. Each section of the fortified palace has its own distinctive feature, be it lattice screens worked in stone to shield the ladies of the harem or the giant open-air ludo board on which the emperor directed his human pieces.

The palace isn't all you can see here either. Just outside the compound is the

Top *Sitting under the archway of the Jodha Bai Palace within the complex* ***Above*** *The grand Jama Masjid, one of the largest mosques in India*

Jama Masjid (the Great Mosque), the burial place of the Sufi saint whose teachings first attracted Akbar to the area. A huge edifice with an imposing entrance, it's a magnificent site in itself.

The village surrounding the site is also worth a stop, mainly for its *nan khatai* biscuits – light, crisp and not too sweet, they go down a treat with a milky cup of Indian tea. After a day wandering around Fatehpur Sikri, these local goodies will be calling your name (ideal, as they come fresh out of local bakeries' ovens in the early evening).

Still want to See the Forbidden City?
If the crowds get a bit much, head to the less congested east and west axis (tour groups usually stick to the central area).

Above *The Anup Talao, a water tank that also acted as an outdoor seating area* ***Right*** *Intricate interiors in the Jama Masjid (the Great Mosque)*

Getting There *Half-hourly buses from Agra take 60–90 minutes. There are also five daily trains from Agra to Fatehpur Sikri.*

www.as.ni.in/fatehpur-sikri

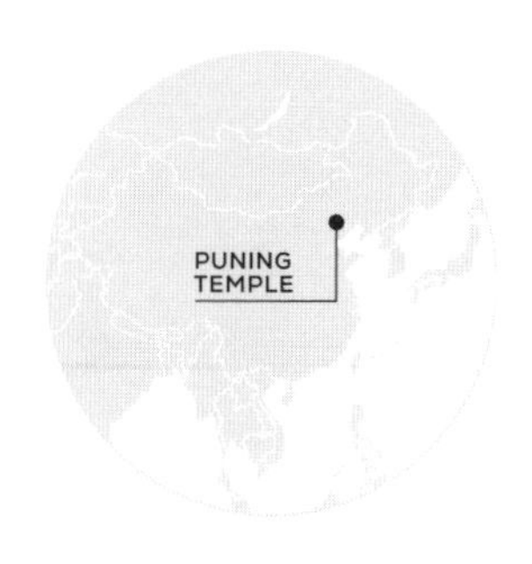

The Temple of Heaven is one of China's most iconic sites but when it comes to ambience, Puning Temple has the edge. Unlike Beijing's famous landmark, this lively complex, surrounded by Chengde's beautiful alpine scenery, is still an active place of worship.

the alternative to *the Temple of Heaven, China*

PUNING TEMPLE

China

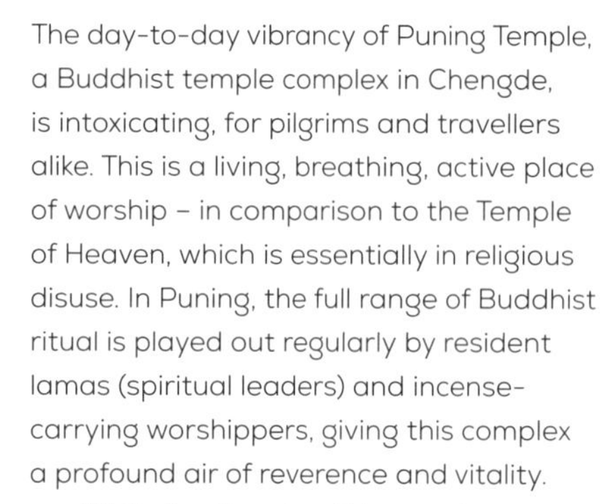

The day-to-day vibrancy of Puning Temple, a Buddhist temple complex in Chengde, is intoxicating, for pilgrims and travellers alike. This is a living, breathing, active place of worship – in comparison to the Temple of Heaven, which is essentially in religious disuse. In Puning, the full range of Buddhist ritual is played out regularly by resident lamas (spiritual leaders) and incense-carrying worshippers, giving this complex a profound air of reverence and vitality.

While the Temple of Heaven is a paradigm of Chinese architectural balance and symbolism (the site is perfectly flat and its buildings are an exercise in restraint), Puning Temple has a spontaneity to it. The complex is a synthesis of Chinese and Tibetan architectural styles, built on the side of a mountain and surrounded by nature. Frequently associated with myths, spirits and magical happenings, mountains in China are characteristic refuges for Buddhist temples. This spectacular backdrop changes colour dramatically from season to season, making the temple a treat to visit any time of the year.

The magnificent temple complex encompasses a series of halls stacked against the slope that serve as sacred portals to the powers of Guanyin, the Buddhist Goddess of Mercy. Her legendary home may be on the island of Putuo Shan, off the Zhejiang coast hundreds of kilo-metres to the southeast, but this UNESCO-protected temple in Chengde is China's most impressive shrine to this deity.

The intangible but boundless energy of Guanyin is represented in the temple's most astounding treasure: a colossal statue standing in the Mahayana Hall, a towering space with several viewing galleries. Standing at over 22 m (72 ft)

Clockwise from left Puning Temple's compact site, dominated by the Mahayana Hall; the huge Guanyin statue; the colourful exteriors of the Puning temple halls

high, the "1,000-arm and 1,000-eye" Guanyin is reputedly the world's largest wooden statue. Gazing up at this giant figure, which sports an eye in each hand, is awe-inspiring to say the least. If you only have time for one Buddhist site in China, this must be it.

Still want to see the Temple of Heaven?
Aim to arrive at the complex at around 7am, when locals often gather to sing, dance and practise tai chi in the temple park. After a wander around the grounds, head to temple, which opens at 8am.

Getting There *The nearest international airport to Chengde is in Beijing. High-speed trains connect Chengde to the capital in less than two hours.*

www.visitourchina.com

Solidly impressive in red-and-white stone, and surrounded by beautiful formal gardens, the Emperor Humayun's fabulous tomb in Delhi was the muse for the later Taj Mahal.

the alternative to *the Taj Mahal, India*

HUMAYUN'S TOMB

India

The magnificent final resting place of the second Mughal emperor, Humayun's Tomb rises spectacularly above its surrounding palm-fringed gardens in Delhi. Finished in 1572, this was the first Mughal garden tomb ever built. It was commissioned by Humayun's loving wife Bega Begum, who oversaw its construction. Some say that it was the inspiration for Agra's later icon, the Taj Mahal, which was constructed some 60 years later by Mughal emperor Shah Jahan to house the tomb of his favourite wife, Mumtaz Mahal. But whereas visitors are faced with a three-hour time limit at the Taj Mahal, at Humayun's Tomb they can drink in the majestic architecture and peaceful greenery at leisure.

Faced in a brilliant colour combination of red sandstone and white marble, Humayun's Tomb stands on a podium exactly in the middle of its spectacular gardens. The intricately carved arches and Persian-inspired bulbous central dome were precursors of monumental Mughal architecture to be found later throughout India. Inside the grand, cool chamber, shielded from the sun by filigree stonework screens, are the silent sarcophagi of Humayun, Bega Begum and over 160 members of the royal family.

The precise geometry of the tomb and its symmetrical gardens, whose design echoes the Qur'an's description of heaven itself, are a visual treat. Walking along the

The beautifully proportioned Humayun's Tomb, surrounded by symmetrical gardens

Arches in the red sandstone façade, inlaid with geometrical patterns of white marble, including the star of Islam

grid of paths that surround the tomb is the best way to admire it from every angle. As you stroll past the tree-lined lawns and water channels, the only sounds you're likely to hear are the gentle chatter of mynah birds and the chirrup of rosy starlings. While you drink in the harmony and geometry, remember the histories of this and Agra's later mausoleum – the first built by a devoted wife for her husband, the second commissioned by an equally devoted husband for his wife.

Still going to the Taj Mahal?
Avoid the foggy winter months and visit between March and June to enjoy the clearest views.

A marble sarcophagus inside Humayun's Tomb, one of the best preserved Mughal monuments in India

Getting There *A taxi from Delhi's Indira Gandhi International Airport takes about 30 minutes to reach the city centre.*

www.humayanstomb.com

Clockwise from top left
Sculptor Korczak Ziolkowski, the designer of the Crazy Horse Memorial; horse sculpture at the memorial; the unfinished rock-cut Crazy Horse pointing over the head of his horse

Deep in the Black Hills of South Dakota, the Crazy Horse Memorial dwarfs Mount Rushmore: if Rushmore's presidents were stacked on top of each other, they'd barely make it halfway up the Sioux warrior's horse.

the alternative to *Mount Rushmore, USA*

CRAZY HORSE MEMORIAL

USA

When Korczak Ziolkowski started work on the world's largest sculpture in 1948, he knew it wouldn't be easy. Offended by the construction of Mount Rushmore in the 1930s – the Black Hills are considered sacred to the Sioux – Lakota Sioux elders had invited the Polish-American sculptor to build an equally grandiose monument to Crazy Horse, a leading Native American figure at the Battle of the Little Bighorn.

The first thing that hits you about the memorial is its huge, almost inconceivable size. The presidents of Rushmore are simply carved onto the rockface, whereas this monument incorporates a whole mountain. Ziolkowski planned to blast, bulldoze and scrape the granite into a 172-m- (563-ft-) high image of Crazy Horse atop his galloping steed, arm pointing over the land he fought for in the 1870s.

Given the gargantuan dimensions, it's perhaps unsurprising that the memorial is still a work in progress. Ziolkowski refused to accept government funds, believing it would compromise the project's mission to preserve the Indigenous cultures of North America, and so construction has become a multi-generational undertaking. The sculptor himself died in 1982, long before the warrior's head was completed in 1998. Even now, it's impossible to estimate when the memorial will actually be finished. No one seems to mind about dates, though. Simply working on the mountain has become almost a spiritual undertaking, as important as the memorial itself.

Still going to Mount Rushmore?
The clear morning light brings the best views; take binoculars to better see the sculpted details and to spot mountain goats clambering around the faces.

Getting There *The nearest airport is Rapid City Regional Airport, which is 65 km (40 miles) away. You'll need a car to visit the memorial.*

crazyhorsememorial.org

NATURAL WONDERS

The wide waterways, tangled jungles and open plains of Western Uganda make up Africa's most biodiverse safari destination, an unspoilt alternative to the more famous Kruger National Park in South Africa.

the alternative to
Kruger National Park, South Africa

WESTERN UGANDA RESERVES

Uganda

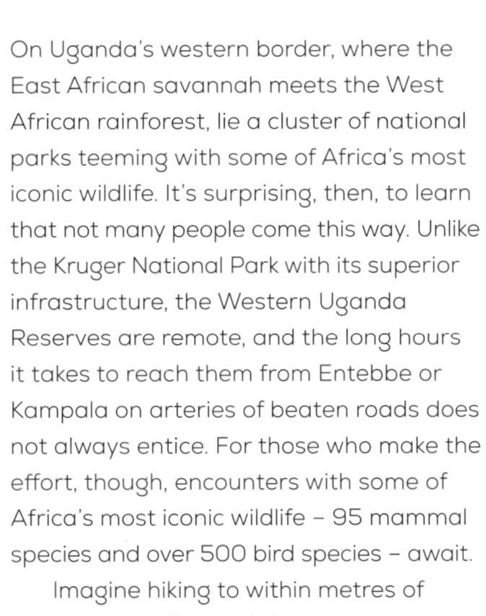

On Uganda's western border, where the East African savannah meets the West African rainforest, lie a cluster of national parks teeming with some of Africa's most iconic wildlife. It's surprising, then, to learn that not many people come this way. Unlike the Kruger National Park with its superior infrastructure, the Western Uganda Reserves are remote, and the long hours it takes to reach them from Entebbe or Kampala on arteries of beaten roads does not always entice. For those who make the effort, though, encounters with some of Africa's most iconic wildlife – 95 mammal species and over 500 bird species – await.

Imagine hiking to within metres of mountain gorillas and chimpanzees, navigating tropical waterways dense with hippos and elephants, and exploring the open plains where lions eye herds of grazing antelopes. Here, it's reality. Bwindi Impenetrable National Park, a sanctuary for almost half of the world's mountain gorillas in the south western region, is almost always the first stop. Up in the mountains, where the air is cool and the rain falls heartily, the vegetation grows thickly, with thin vines snaking on the trees trailing the rainforest's muddy floor. It's a hard hike to the gorillas and sightings aren't guaranteed. But listen: was that rustle the sound of one passing by?

Neighbouring Bwindi, the Mgahinga Gorilla National Park is a quieter, smaller gorilla sanctuary covering the plush slopes of the three Virunga Volcanoes. It may be small, yet this park is mighty. Trekking to the 100 or so gorillas is the highlight for many, but hiking the mountains of Gahinga, Sabyinyo and Muhavura offers a chance to glimpse yet more wildlife (golden cats, leopards, duiker) along the luscious slopes.

A small herd of giraffes roaming the vibrant grasses at Murchison Falls National Park

Above A gorilla nestled among forest vegetation at Bwindi Imprenetrable National Park

Enchanting scenery beckons further along the western trail at the Queen Elizabeth National Park, on the Rift Valley floor. It certainly wears the crown when it comes to biodiversity, with savannah, forest and aquatic habitats converging and extending from Lake George in the northeast to Lake Edward in the southeast. Connecting the two lakes is the Kazinga Channel – ideal for a water safari, where hippos poke their weighty foreheads out among wayward crocodiles as your boat glides through the water. At the remote Ishasha Plains in the south, where acacias and cactus trees dominate the landscape, the famed tree-climbing lions lazily digest the day's meal.

Western Uganda continues to surprise with its terrain, especially at the Rwenzori Mountains National Park. This alpine blend of grasslands and immortal snow-capped mountains attracts a sweet variety of mammals and birdlife, who join you on the mammoth treks through the park. If all those twitching birds have got you itching for more, continue north to Uganda's youngest treasure: Semuliki National Park. Sitting in the Albertine Rift Valley between Lake Edward and Lake Albert, it's home to unique bird species found nowhere else in the country – the crested marimba, yellow-throated cuckoo and the tiny African piculet, for starters.

The last stop on the western trail is by no means least. North of the Albertine Rift Valley, the Murchison Falls National Park is Uganda's largest, providing a home for 76 mammal species and hundreds of birds. Boats follow the Nile to the base of Murchison Falls, where thousands of litres of water crush through a gorge

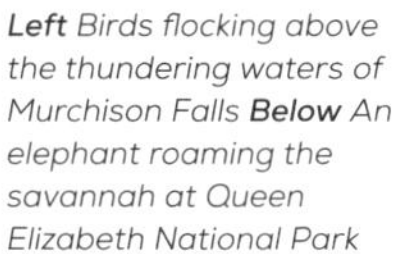

Left *Birds flocking above the thundering waters of Murchison Falls* ***Below*** *An elephant roaming the savannah at Queen Elizabeth National Park*

and fall dramatically into Lake Albert. Lions, giraffes, elephants, buffalos, Uganda kobs and oribi inhabit the palm-studded grassland nearby, making for an electrifying game drive experience to end your trip.

Still going to Kruger National Park?

Kruger is a wildlife-lover's paradise: South Africa's largest national park, it's got a mammal checklist of 150 species. The south region of the park is most popular, being more developed for tourists, but the north regions from Punda Maria are just as bountiful (without the clogged roads and crowded camps).

Getting There *Most visitors explore the western parks by road over one or two weeks. Private safaris can be arranged through operators in Kampala.*

www.ugandaparks.com

MORE LIKE THIS

LOPÉ NATIONAL PARK
Gabon

This UNESCO World Heritage Site is home to savannah grasslands that were created over 12,000 years ago. The park is vital for research and is home to western gorillas, sun-tailed guenon, chimpanzees, black colobus and forest elephants.

SAMBURU NATIONAL RESERVE
Kenya

Wildlife fanatic? You're in for a treat. The rare species of Grevy's zebra, Somali ostrich, reticulated giraffe, gerenuk and the Beisa oryx – collectively known as the Samburu Special – inhabit this reserve.

LUANGWA VALLEY
Zambia

This area is said to be the honey-pot of Zambia for its lush vegetation and rich wildlife. The valley is home to four national parks: South Luangwa, North Luangwa, Luambe and Lukusuzi, which are home to the Big Five and Thornicroft's giraffe.

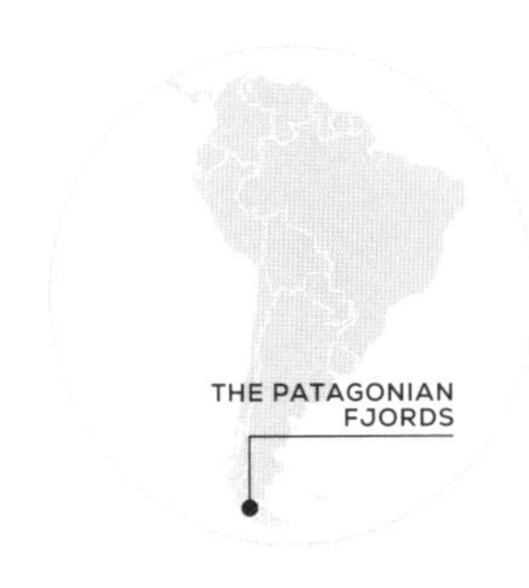

A magical world of water and ice, Patagonia's fjords, rising from the depths of the South Pacific, eclipse Norway's famous waterways with their wild and untouched beauty.

the alternative to *the Norwegian fjords, Norway*

THE PATAGONIAN FJORDS

Chile

Ice floes littering the waters in the spectacular Calvo Fjord, on the edge of the Sarmiento Channel

At the bottom of the earth lies a dazzling world of hanging glaciers, floating icebergs, plunging waterfalls and towering peaks. These haunting wonders receive just 12,000 visitors a year – a fraction of the number that descends on their Norwegian rivals.

The Patagonian fjords experience begins in the city of Punta Arenas in Chile, from where cruises cross the legendary Magellan Strait (Estrecho de Magallanes) and float south to Cape Horn through a labyrinthine network of channels, islets and inlets. As you sail, mountains rise out of the sea and glacier after glacier cascades down the steep rock faces. Boat excursions magnify the thrilling drama at water level – approaching Glaciar Pía, visitors watch giant blocks of turquoise ice fall from the glacier's face into the Pía Fjord in a concerto of thunderous cracks and booms.

Small-boat trips bring you incredibly close to Chile's marine wildlife. Dolphin pods play in the icy waters here while giant elephant seals and smaller sea lions slumber in the sheltered coves. Isolated beaches are also home to thousands of cute Magellanic penguins.

After four days of navigating these sheltered waters, boats reach desolate Cape Horn. Here, the Atlantic and Pacific

Magellanic penguins gathering on a rock in the Patagonian snowy landscape

oceans collide, ice floes pepper the sea and the howling winds that once thwarted expeditions lash against steep, ragged cliffs. It's not difficult to believe that the ends of the earth are within reach.

Still want to see the Norwegian fjords?
Having fallen victim to their own beauty, Norway's UNESCO World Heritage fjords welcome scores of tourists every year. To find a peaceful pocket, head for the less-visited Naeroyfjord and Lysefjord. The latter is overlooked by the 600-m- (1,970-ft-) high Pulpit Rock.

Getting There *There are daily flights from Santiago to Punta Arenas in southern Chile, the departure point for cruises to the Patagonian fjords.*

www.australis.com

KAIETEUR FALLS

Hidden away from the crowds and surrounded by the verdant Amazon Rainforest, Kaieteur Falls are a staggering four times higher than Niagara Falls.

the alternative to *Niagara Falls, Canada*

KAIETEUR FALLS

Guyana

Tumbling at a ferocious speed of 663 cubic m (23,414 cubic ft) per second, Guyana's legendary waterfall is one of the most powerful in the world. Its scale, however, is in no way reflected by its visitor numbers – a mere 40,000 a year. Buried deep in Guyana's jungle, and protected by one of South America's largest tracts of undisturbed rainforest, Kaieteur Falls aren't the type of natural wonder you'll be competing for a view of; here, in this unique realm of wilderness, you'll likely meet more wildlife than tourists.

Still want to see Niagara Falls?
You can't escape the crowds but you can get up close by taking the *Maid of the Mist* boat trip to the base of the falls.

Getting There *There is no road access. Visit either by plane from Georgetown (an hour's flight), or on foot via a five-day wilderness trek.*

www.guyanatourism.com

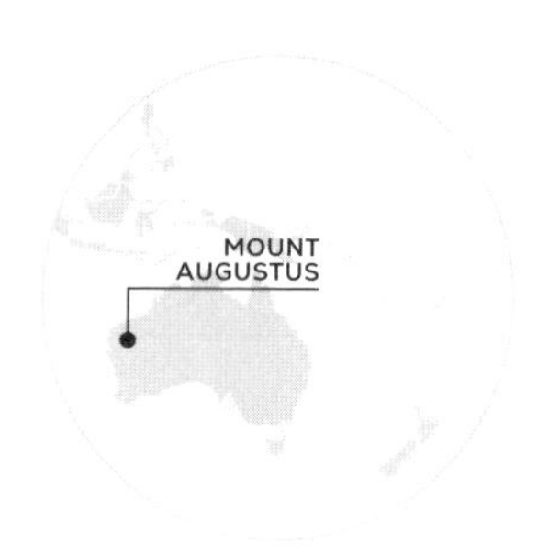

Known as Burringurrah to the Wajarri Aboriginal peoples, Mount Augustus is twice the size of Uluṟu, and ten times as remote. If you're up for a challenge, head up to the top of this spectacular monolith.

the alternative to *Uluṟu, Australia*

MOUNT AUGUSTUS

Australia

The spectacular solitary peak of Mount Augustus rises 715 m (2,346 ft) from the ochre desert floor. While this hulk of sedimentary rock may look like its more famous cousin Uluṟu, it's in fact double the size.

To the Wajarri peoples, this stately mountain is named Burringurrah – after a Dreamtime figure who turned to stone after trespassing. Modern-day visitors won't meet this fate, however, as, unlike at Uluṟu, where Aboriginal customs prohibit climbing, to reach Burringurrah's summit, you're limited only by your fitness – allow eight hours for the gruelling 12-km (8-mile) path. Gentler walks circle the mountain, with many trails hemmed by wattle shrubs and cassia flowers that blush in springtime. Walkers may even spot emus, red kangaroos and small rodents as they wander.

There are also a number of rock art panels near the mountain. A 49-km (30-mile) driving track, which loops around Mount Augustus, passes the visitor sights of Beedoboondu, Ooramboo and Mundee – all of which are home to ancient rock carvings by the local Wajarri peoples.

Vast Mount Augustus, glowing red at sundown

Still going to Uluṟu?
Stay at Longitude 131 *(www.longitude131.com.au)*. The camp has an Aboriginal artist in residence and has been landscaped for minimal environmental impact.

Getting There *You'll need a few days in a 4WD to drive from Carnarvon (470 km/292 miles) or Meekatharra (354 km/220 miles) in Western Australia to Mount Augustus.*

www.parks.dpaw.wa.gov.au

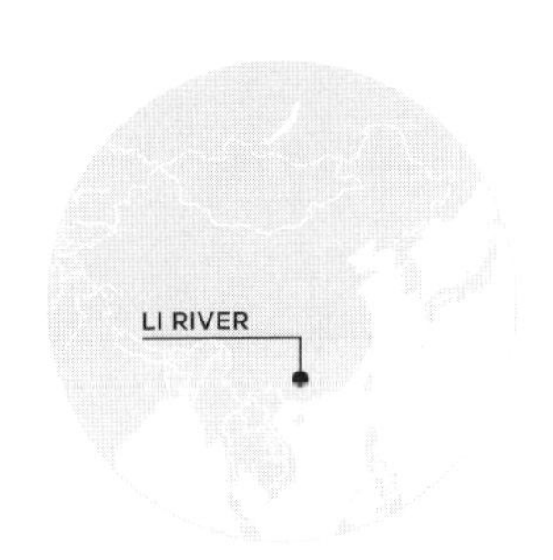

Visitors intrigued by the geology of Halong Bay would do better visiting China's Li River, which has similarly distinctive limestone mounds rising from its waters, but far fewer tourist boats interrupting the view.

the alternative to *Halong Bay, Vietnam*

LI RIVER

China

Flowing through the upper section of Guanxi's Pearl River Basin, the Li River snakes through bulbous limestone hills and towering peaks, many of which are cloaked in emerald forests. It's a landscape similar to Vietnam's popular Halong Bay, but it's even more mesmerizing.

Over the millennia, the river has carved huge archways into the rocks here, the most famous of which is Elephant Trunk Hill (aptly named for its resemblance to an elephant drinking from the river). Beneath the ground, yawning caves are so grand they have a church-like atmosphere, with stalactites and stalagmites acting as the earth's own form of stone masonry.

Cruises along the Li River also give a glimpse into rural China: water buffalo bathe in the shallows while cormorant fishermen work in tandem with their birds. The pace of life is slow – there's always movement, but never enough to disturb the tranquillity of the scenery, or the cruise.

Still going to Halong Bay?

Opt for kayaking instead of a cruise – it gives you much more flexibility, so you can get closer to the rock formations.

Elephant Trunk archway, one of many formations carved into limestone rock by the Li River

Getting There *The most scenic cruise is from Guilin to Yangshuo. Guilin has an airport (with flights to Hong Kong and Beijing) and a train station offering high-speed trains to major cities.*

MORE LIKE THIS

RAJA AMPAT
Indonesia
Raja Ampat, which means "four kings", combines verdant green islands with turquoise waters and some of the world's richest marine biology.

WULINGYUAN
China
Wulingyuan is Halong Bay but without the water: more than 3,000 rock formations rise in tree-topped towers from a gorge.

Italy's magnificent lakes are synonymous with beauty and luxury, but when it comes to sheer natural splendour, they're no match for Chile's spectacular volcanic wonders. Take a trip off the tourist path to discover this tantalizing landscape.

***the alternative to** the Italian Lakes, Italy*

THE CHILEAN LAKES

Chile

Beautiful though the lakes of northern Italy are, ringed by tall mountains and overlooked by elegant villas and medieval castles, their allure pales when compared to the dramatic lakes of southern Chile. Travel to the very edge of Patagonia and you'll discover this fairy-tale region of shimmering blue water, emerald forests, looming Andean peaks and steaming hot springs that bubble under smouldering volcanoes. Italy's lakes may have inspired many a Romantic poet, but Chile's land of fire and water has been a source of legend for millennia.

Rare primordial beauty is everywhere you turn in Chile's Lake District, which is anchored by two cities: to the north lies Temuco, to the south Puerto Montt. The lakes are, as the name suggests, the star of the show: they spread across the whole region, their mineral-rich, azure waters lapping serenely at the feet of volcanoes. The shores are just as dramatic, frosted with a combination of black volcanic powder and fine white sand – often too tempting not to walk across barefoot.

Standing in this untamed wilderness may give you a sense of the elemental, but just wait: you're only at ground level. A climb up the volcanic slopes that over-look the lakes promises even more wonder. Roads here lead to mountain spas, where travellers bathe in hot springs that bubble up from beneath the earth's crust. Horse rides on these slopes reveal lush virgin forests bursting with secret waterfalls, gushing streams and ancient groves of monkey-puzzle trees. For hikers eager for a challenge, trails lead to the summits of the great Villarrica and Osorno volcanoes – sacred sights to the Mapuche, the native people who've lived in the Lake District for thousands of years. Venturing up here, where craters glow with molten lava and spit gases into the atmosphere, promises a glimpse at the earth's fiery inner layers.

An iconic Lake District view: the Puntiagudo-Cordón Cenizos volcano rising above Todos los Santos Lake

Peppered throughout this scenic region are a number of picturesque towns – which make for romantic stopovers while travelling around the lakes. Many of them are home to historic wooden churches, built by German colonists in the 19th century. Perhaps more enticing for weary travellers, however, is the promise of delicious regional food and quality Chilean wine – both a worthy match for Italy's culinary treasures.

Still want to see the Italian Lakes?

Being easily accessible comes with a cost: the Italian Lakes swarm with tourists during high season (Jun–Sep), when restaurant prices rise higher than the peaks. To escape the crowds, visit either side of summer and consider heading to a smaller destination, such as Lake Iseo, instead of the famous Lake Garda.

Getting There *Flights to Chile's Lake District depart daily from Santiago to Temuco's airport. From here you can either rent a car or travel by bus.*

www.chile.travel

Top *Travellers taking a break above the roaring Petrohué Waterfalls, one of the Lake District's most impressive waterfalls* ***Above*** *Seals basking on the beach in Puerto Montt – the port city that marks the southern tip of the lakes region*

Skip the popular Grand Canyon in Arizona in favour of one of the remotest spots on the planet – Fish River Canyon, hidden in the deserts of Namibia. Africa's greatest natural wonder is second in size only to its famous cousin but offers a beguiling isolated beauty.

the alternative to *the Grand Canyon, USA*

FISH RIVER CANYON

Namibia

Arizona's Grand Canyon is the largest and most famous in the world, but in southern Namibia there's a spectacular canyon that's almost as vast – the Fish River Canyon. Situated just north of the border with South Africa, between the Namib and Kalahari deserts, the magnificent Fish River Canyon is reached by roads across huge expanses of barren landscapes, dotted with only the occasional cluster of desert plants and cacti. As a result of this physical isolation, it's very likely that you'll have breathtaking views all to yourself.

The canyon was created by the wear and tear of the Fish River (Namibia's longest interior river) and the collapse of the valley bottom through a shift in the earth's crust some 500 million years ago. It measures 160 km (100 miles) in length and up to 27 km (16 miles) in width. The jaw-dropping inner canyon is 550 m (1,800 ft) deep. Multiple lookouts along its rim give views over the canyon but the best way to get to know it is on a five-day 85-km (53-mile) hiking trail, which crosses half of the canyon's length. This can certainly be tough: there are no facilities and all belongings must be carried with you. Winter (June to September) is the best time to visit, when the temperature is much cooler and the weather drier than in the extremely hot summer months.

Starting at the Hobas viewpoint at the northern end, the trail descends down a very steep path, and follows the sandy river bed as it snakes between the towering canyon walls all the way to the Ai-Ais Hotsprings. As you hike along kilometres of chestnut-coloured sand, past expanses of red and orange rock

Walking along the river bed through the magnificent canyon landscape

formations and sparse desert bushes, you'll feel tiny in this titanic ravine. The only sounds are the cries of abundant birdlife. Fish eagles, pied kingfishers, Goliath herons and countless other birds circle above. As for willdlife, most hikers will come across wild horses, zebras, baboons, springbok, klipsringer or kudu at some point along the route.

At the canyon's southern border, the Ai-Ais Hotsprings mark the end of the trail. There's no more satisfying end than soaking weary limbs in its natural healing waters, rich in chloride, fluoride and sulphur, as you gaze up at the towering canyon peaks around you.

Still going to the Grand Canyon?
Make your way to Shoshone Point, a (usually) quieter outlook with stunning views, just a short walk down a dirt road from the north side of Desert View Drive.

The spectacular rock formations of the gigantic Fish River Canyon, carved out by the Fish River

Getting There From Namibia's Hosea Kutako International Airport in Windhoek it's an eight-hour drive to Fish River Canyon. Several companies organize tours.

www.namibiatourism.com

MORE LIKE THIS

WAIMEA CANYON
USA
Hawaii's "Grand Canyon of the Pacific", which almost cuts the verdant valley of Kaua'i in half, holds an amazing, ever-changing panoply of colours. The views out over the ocean are magnificent.

COPPER CANYON
Mexico
Northern Mexico's Copper Canyon is a league of six distinct canyons whose rock walls are a green/copper colour. Pine and oak trees abound, but it's in October, when wild flowers blossom, that the landscape really comes into its own.

COAL MINE CANYON
USA
Located in the remote desert of northeastern Arizona, Coal Mine Canyon sits on the border between the Hopi and Navajo reservations. It's the stunning rainbow colours, spiral shapes and textures of the sandstone rock face that are the draw here.

The Amazon promises an epic adventure, there's no doubt about it. Wildlife-spotting opportunities, though? For that, make your way to the Orinoco, a river teeming with thousands of plants and animals.

the alternative to *the Amazon, South America*

THE ORINOCO

Venezuela

In the language of the Indigenous Warao peoples, the Orinoco means "a place to paddle" – and it's certainly just that. This vast river gives you more opportunity for up-close access to South America's astonishing wildlife and landscapes than the more famous Amazon. Yes, the Amazon bursts with biological richness, but much of the river is either inaccessible or very expensive to reach; the Orinoco, on the other hand, is easier to get to and is navigable for most of its length.

Arcing across Venezuela past Colombia, the mighty Orinoco river system runs a breathtaking topographical gamut, from the tall Parima Mountains bordering Brazil, through tropical forests, humid *llanos* (seasonally flooded plains) and marshy deltas and mangroves, before finally emptying into the Atlantic. It has a mind-boggling diversity

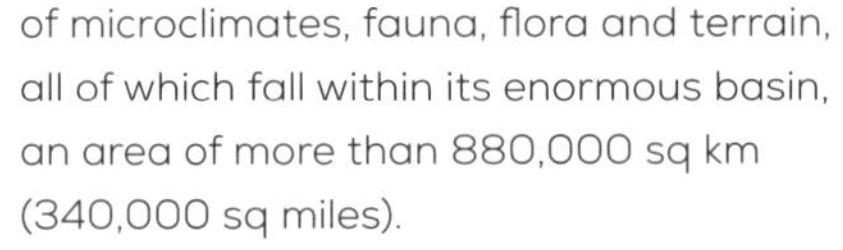

of microclimates, fauna, flora and terrain, all of which fall within its enormous basin, an area of more than 880,000 sq km (340,000 sq miles).

Here you'll find one of the last pristine ecosystems on the planet (though this might not last for long as it's increasingly under threat from deforestation). Exciting new species are discovered regularly. At last count, this still largely uncharted river was home to more than 17,000 plant species, 1,400 bird species, 1,200 fish species, and at least 340 different types of mammal. Keep your eyes peeled and you're likely to spy alligators, pink dolphins, boa constrictors, herons,

A green aracari toucan, which can be spotted in the lowland forests along the Orinoco River

toucans, howler and capuchin monkeys, and pumas – and that's more you can say for the Amazon, where foliage is so dense that it's hard to see much of anything at all. For one of the best places to see this abundant wildlife, head to the Orinoco Delta, the swamp forest at the river's mouth. It's thought by many to be the richest area in aquatic life in the whole of South America. It's time to pick up that paddle: you won't find this kind of diversity anywhere else on earth.

Still want to explore the Amazon?

There's no doubt that the Amazon, by dint of its sheer size, is an awesome sight. For those who want to visit South America's longest river, one option is to fly to Manaus in Brazil, and start the journey there. The city's central location and numerous operators will make your planning easier.

Gliding in a small boat through lush forest on the Orinoco Delta at the river's mouth

Getting There *Fly to Manuel Carlos Piar Guayana Airport in Ciudad Guaynana, Venezuela to start your trip at the mouth of the Orinoco.*

RÍO PARAGUAY
South America
The Río Paraguay, another legendary South American river and gateway to Brazil's Pantanal and Mato Grosso ecosystems, makes for a fascinating alternative for those who want a little less jungle and a lot more open space.

HUANG HO
China
For stunning mountain scenery and a chance to see ancient and modern civilizations side by side, head for China's lifeline, the Huang Ho. The river runs east for 5,500 km (3,400 miles) from the Bayan Har Mountains in western China.

ALLAGASH RIVER
USA
This 150-km (95-mile) waterway in the northeast USA might be comparatively short, but it's thrilling to explore. With its rapids, black water and deceptively placid appearance, the Allagash commands respect at every bend.

While Iceland's natural wonders may steal the spotlight, Bolivia is the Narnia of the geothermal world. Far off the beaten track and hidden from the crowds, the fantastical geyser field of Sol de Mañana, and the landscape surrounding it, is one for the ages.

***the alternative to** Iceland*

SOL DE MAÑANA

Bolivia

Iceland may feature some of the world's greatest geothermal activity, but this little island is feeling the pressure (and not just from under the earth). Crowds come hand in hand with natural wonders here, but high in the hinterland of southwest Bolivia, it's an entirely different world all together.

The Sol de Mañana geyser field is a frenzy of geothermal force. Pressurized jets of scalding hot steam escape from fissures in the earth's crust, fizzing into the air at heights of up to 15 m (49 ft). Meanwhile, mud pots bubble in a myriad of colours, from sky blue to murky grey and blood red. These devilish pools are scattered across a vast plateau, creating vivid dashes of colour against the landscape's ochre-hued rock. On the horizon, a line of perfectly conical stratovolcanoes adds to the otherworldly appearance, as if you've somehow wound up on Mars rather than earth.

Located miles from civilization, Sol de Mañana only welcomes around 140,000 tourists every year – a fraction of those who whistle-stop tour around Iceland's volcanic landscape. Most visit this wild, geothermal area on a tour of the region's natural wonders, which begins in the city of Uyuni. Among the tour's other ethereal stops are Laguna Colorada, a lake whose waters glow an intense algae-induced red, in contrast to the white borax islands and the lustrous pink of the flamingos that stalk its salty waters. Nearby, on a high plateau where vicuñas (wild camelids) roam, a series of rocks buffed by the wind and sand rise from the desert in bizarre shapes – Dalí and his Surrealist pals would have a field day here.

Tours also stop at Salar de Uyuni, just further north. Here lie the world's largest salt flats, spanning the earth with crisp, snow-white salt. Laid out in a naturally

Clockwise from top right *The steamy Sol de Mañana geyser field; driving along the dazzling salt flats of Salar De Uyuni; flamingos in the Laguna Colorada*

occurring series of hexagonal tiles, the landscape mimics the beginnings of a fantastical board game played by giants. As at Sol de Mañana, you can expect to find yourself here alone, staring at a landscape so unearthly you'll ask yourself later if it ever really existed at all. Well, the answer is it certainly does, and it's more than a match for Iceland.

Still want to go to Iceland?
As Iceland barely registers full darkness during the summer months, consider visiting the most famed geysers, such as Strokkur, on a June or July night.

Getting There *Visitors can fly from La Paz (Bolivia's capital) to Uyuni in less than an hour. Trains and buses are also an option.*

www.salardeuyuni.com

An untouched world filled with endemic species, the Mindo cloudforests feel like their Costa Rican counterparts did before they rose to fame. These lesser-known forests also promise a wealth of fun-filled activities and, most importantly, delicious local chocolate.

the alternative to *the Monteverde cloudforests, Costa Rica*

THE MINDO CLOUDFORESTS

Ecuador

Tucked away in Ecuador, one of the most biodiverse countries in the world, lies the extraordinary Mindo-Nambillo Cloud Forest. It's one of those places that will leave you doubly astonished, by its beauty and by the fact that it isn't more widely known – though that's changing fast. Covering nearly 20,000 hectares (49,000 acres), the reserve has a spectacular wealth of native flora, including some of the world's rarest orchids, and fauna, particularly tropical butterflies and birds – many of which were once thought to be extinct but actually managed to survive in this tropical paradise. Known as *bosques nubosos* in Spanish, these clouds forests are also smothered in lush vegetation: mosses, lichen, ferns, creepers, bromeliads and epiphytes wrap themselves around tree trunks, while swirling clouds of mist shroud the canopy.

Clouds rolling over a lush valley in the verdant Mindo-Nambillo Cloud Forest reserve

Perched on the western slope of the Pichincha volcano, less than 100 km (62 miles) by road from the capital Quito, the small, subtropical town of Mindo is the gateway to this prime slice of Ecuador's cloud forests. Although the *bosques nubosos* here are far from undiscovered – they are a popular destination for Quito residents seeking a rural break – they receive fewer visitors than their equivalents in Costa Rica, whose trails can often feel overcrowded. Unlike forests elsewhere, there are also stringent regulations here: you won't find mass tour guides, nor five-star luxury hotels that undermine the area's ecology. The government established a buffer zone in the park, the first of its kind in South America, where visitors can lodge and stock up on supplies. Guides are only allowed to set up shop if they pass extensive tests.

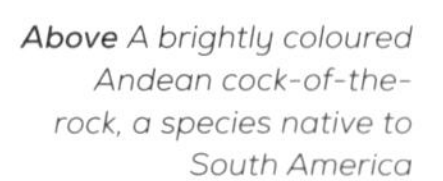

Above *A brightly coloured Andean cock-of-the-rock, a species native to South America*

Above *A chair lift gliding through the tropical rainforest (an ideal way to spot bird life)* ***Right*** *Sunrise over the cloud forests*

The cloudforests around Mindo are best known for their profusion of bird-life. More than 500 species have been spotted in the region, including scores of different hummingbirds and the rare long-wattled umbrellabird, which has, as its name suggests, a long dangling wattle and a quiff-like tuft on its head. The area's most famous avian resident is the charismatic Andean cock-of-the-rock. Easily identifiable thanks to their striking red or orange plumage, the males perform distinctive "dances" and make piercing squeals to challenge rivals and attract a mate. Watching these captivating courtships is a mesmerizing experience. It's not hard to spot birds in Mindo, but to maximize your chances of encountering these more unusual species it's worth hiring the services of an experienced local guide.

Within these magical forests, you can also glimpse countless troupes of howler, capuchin and spider monkeys, as well as prowling big cats such as pumas and ocelots. If you're very lucky, you may even spot a reclusive Andean spectacled bear – the creature that inspired a certain duffel coat-wearing, marmalade-loving, star of page and screen: Paddington Bear.

Beyond bird- and wildlife-watching, there's a huge range of activities available around Mindo, including ziplines, mountain bike routes and white-water rafting trips. If you need an energy boost after all that physical activity, artisan chocolate producers promise delicious treats from locally grown cacao – which, many argue, is more flavourful than Costa Rican blends.

History fans will also be happy to hear that Mindo's wider region features numerous archaeological sites dating back to the Inca and pre-Inca cultures. The latter include a ceremonial complex built by the

Below One of the many waterfalls that cascade through the cloud forests

Yumbos, a society that emerged around 800 BC, at the village of Tulipe. Nearby, are the ruins of a set of sunken pools, which Yumbos astronomers are believed to have used as giant mirrors to study the movements of the sun, moon and stars. These fascinating sites highlight the sheer diversity of the cloudforests around Mindo – and act as a reminder that there is still much more to be discovered in this dazzling region.

Still going to Monteverde?

Monteverde's cloudforests may draw a steady stream of travellers, but they are still beautiful places to explore. To see the cloudforests with fewer crowds, head to areas that are less heavily touristed such as the Santa Elena Reserve.

***Getting There** Buses link Mindo, in the Pichincha province, to Quito in about two hours. You can also hire a taxi from Quito's international airport.*

www.ecuador.travel

MORE LIKE THIS

BORNEO'S JUNGLE
Brunei, Indonesia and Malaysia
The rainforests on Borneo (which cover the majority of the island) are still among the least explored and the most biodiverse on the planet.

MATA ATLANTICA
Brazil
This World Biosphere Reserve holds more than 40 per cent of the world's known tree species. It also has its fair share of rare plants and wildlife.

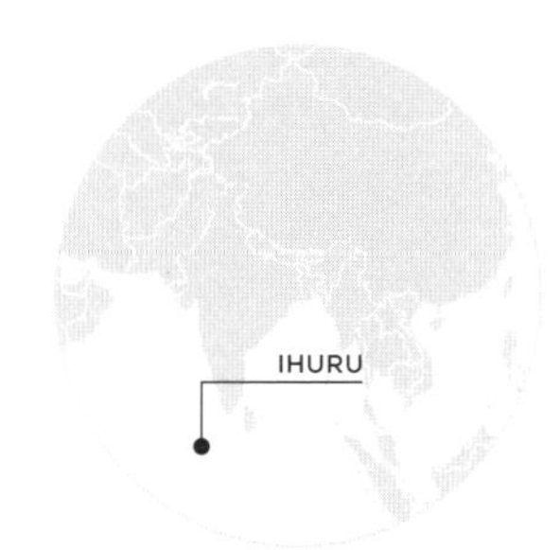

They may have filmed The Beach *on Ko Phi Phi, but it can't be a fantasy island getaway if you have to share it with a crowd. Ihuru is the real desert-island deal – a slice of peaceful paradise nestled in the Maldives.*

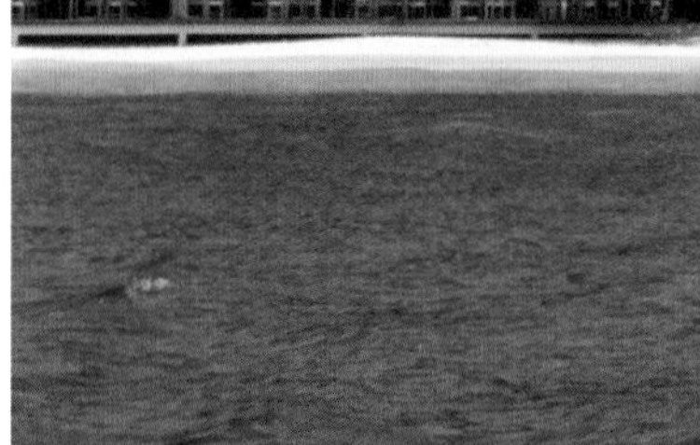

***the alternative to** Ko Phi Phi, Thailand*

IHURU

Maldives

There are four essential ingredients for an island retreat – sand, sea, sun and seclusion – and Ihuru has them all. This is a proper desert island: the only noise comes from surf breaking over the distant reefs and the rustle of palm fronds, and the only reason to stir is the incoming tide. Thailand's Ko Phi Phi has the surf and the palm fronds, of course, but the constant flow of tour boats is the nautical equivalent of rush hour in downtown Bangkok. When you close your eyes and picture the perfect island paradise, it's Ihuru you see.

Formed from the tip of a submerged volcano, Ihuru is not so much an island as a beach with shade. Set in its own mini-atoll, the island is a single copse of palm trees, ringed by icing-sugar sand. Ihuru does not have beaches (plural) – the whole island is a beach, fringed by a turquoise lagoon of warm water and encircled by a sheltering curtain wall of reefs. You may well want to explore its breathtaking undersea world, or you might just prefer to doze beneath a palm tree or stroll along the sand. There's no pressure – which is rather the point of coming here.

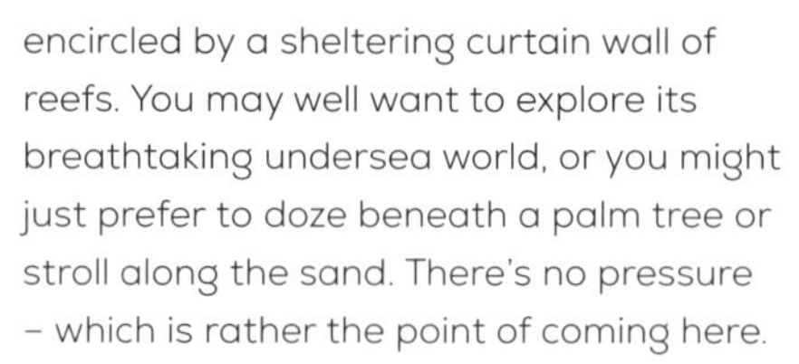

Under the Maldives' unique model for tourism, every developed island is home to a single, self-contained resort. Unlike Ko Phi Phi, where resorts spring up like mushrooms after April showers, accommodation on Ihuru is as private and secluded as you can get. Basking beneath the palm trees, the swish villas of the Angsana Ihuru resort – the only resort you'll find on the island – are

Aerial view of Ihuru and neighbouring Vabbinfaru islands, circled by coral reefs

The palm tree-lined sandy beach and crystal-clear waters of Ihuru, an island paradise

a favourite of the fashion set, who come here to be pummelled or pampered by expert practitioners in the spa, or left alone to rock gently on a Maldivian swing in front of an empty stretch of sand.

Ihuru is luxurious, certainly, but the resort buildings and villas are scattered like fallen coconuts between the palms, and the first footprints on the beach every morning will probably be your own. The whole purpose of an island getaway is being able to get away from it all – who wants to get away from it all with everybody else?

Still want to sunbathe on Ko Phi Phi?

We get it: Ko Phi Phi is beautiful. If you can't resist the urge to see it, head to the smaller, more intimate bays and secluded resorts on Hat Laem Thong and Ao Lo Bakao, rather than the heavily developed beaches along the sandy central isthmus of the island.

Getting There *Fly to Malé International Airport on Hulhule island. From here it's a 20-minute speedboat ride to Ihuru.*

www.visitmaldives.com

MORE LIKE THIS

RADHANAGAR BEACH, HAVELOCK ISLAND, ANDAMAN ISLANDS
India

Radhanagar is a pristine, tree-fringed curl of sand on Havelock island, a tiny dot of land in the middle of the Bay of Bengal. The waters offshore are rich in marine life.

GOLDEN BEACH
North Cyprus

Little has changed in North Cyprus since the island was divided in 1974. More turtle tracks than footprints mark the sands on this 17-km (10-mile) stretch of beach on the remote Karpas peninsula, which you're likely to have all to yourself.

SANDAY, ORKNEY
UK

Imagine an island 2 km (1 mile) wide by 20 km (12 miles) long, encircled by white sand and the warm waters of the Gulf Stream – in Scotland, unexpectedly. Remote Sanday comes into its own in summer, when pleasant weather graces the beaches.

Located off Brazil's northeast coast, the archipelago of Fernando de Noronha is a near mirror-image of the Galápagos to the west of Ecuador. These untamed islands will make you feel like you've stepped back in time into pristine nature.

the alternative to *the Galapágos Islands, Ecuador*

FERNANDO DE NORONHA

Brazil

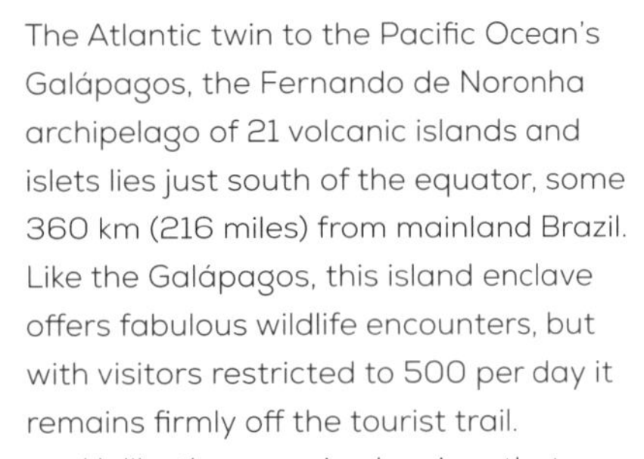

The Atlantic twin to the Pacific Ocean's Galápagos, the Fernando de Noronha archipelago of 21 volcanic islands and islets lies just south of the equator, some 360 km (216 miles) from mainland Brazil. Like the Galápagos, this island enclave offers fabulous wildlife encounters, but with visitors restricted to 500 per day it remains firmly off the tourist trail.

Unlike the organized cruises that predominate at the Galápagos, the perfect way to enjoy Fernando de Noronha is to settle into one of the small *pousadas* (hotels) on the main and only inhabited island, Ilha Fernando de Noronha. From here, you can make forays by wooden fishing boat to neighbouring isles, ride a dune buggy around pristine beaches or explore the trails and dirt roads on foot or by mountain bike.

Everywhere you turn, you'll encounter glorious scenery. Rugged pinnacles pierce the sky and sheer-faced cliffs spill onto idyllic strands such as Praia do Leáo, where marine turtles crawl ashore to lay their eggs December to May. Migratory birds flock in their teeming thousands, joining permanent residents, including the magnificent frigatebird, red-billed tropicbirds and the endearing brown, masked, red-footed and blue-footed boobies. In fact, Fernando de Noronha is home to the largest concentration of tropical seabirds in the western Atlantic.

And the Baía dos Golfinhos has the densest population of resident dolphins in the world. If you time it right and head out just after dawn, you should be lucky enough to see spinner dolphins and even whales frolicking close to shore. The crystal-clear waters of the islands teem

The turquoise waters, picturesque rocks and numerous islets of the paradisiacal Fernando de Noronha

A spinner dolphin, the most common dolphin species seen around the archipelago

with other marine life too. Giant shoals of brightly coloured fish dart around, rays and skates glide over the sandy sea floor, and turtles paddle under the waves. A top pick for snorkelling is Praia do Atalaia, which has some of the shallowest and calmest waters. But that's only one of many turquoise natural pools backed by lush foliage in Fernando de Noronha. Here, tropical paradise is all around.

Still going to the Galápagos?
Take tours in small charter boats led by local guides. Stick to marked trails, and don't touch or feed the wildlife.

Getting There *The Aeroporto Fernando de Noronha is served by flights from Natal and Recife on mainland Brazil.*

www.noronha.pe.gov.br

MOUNT RORAIMA

Table Mountain may be so-named but it's far from being the world's only flat-topped summit. Among the others, Mount Roraima stands out for some of the oldest rock formations on earth.

the alternative to
Table Mountain, South Africa

MOUNT RORAIMA

Venezuela

Situated in Venezuela's Canaima National Park, Mount Roraima is the highest of the table-top mountains or *tepuis* ("houses of the Gods", in the language of the Pemón peoples) in the Guiana Highlands of South America. Guided hikes take the intrepid up a steep rugged trail to the flat top, which covers a whopping 30-sq-km (11-sq-mile) area. From here there are spectacular views of Venezuela, Guyana and Brazil (the mountain straddles the border at which these countries meet). A hike across the pleateau reveals stunning waterfalls, rocks with interesting shapes formed two billion years ago and ancient insect-eating plants. It's a veritable Jurassic wonderland.

Heading up Table Mountain anyway?
First check the weather forecast. Cloud cover can obstruct views and high winds stop cable cars running.

Getting There *From Venezuela's Santa Elena de Uairén Airport catch a bus to Paratepui, where you can hire a guide.*

https://national-parks.org/venezuela/canaima

Keen to see the rusty tones of autumn in all their splendour? New England's fall colours are a sight to behold but over the border, the Laurentian Mountains put on a foliage show that's just as stunning.

the alternative to *fall in New England, USA*

FALL IN THE LAURENTIAN MOUNTAINS

Canada

The forests of New England may be your first port of call for a colourful autumnal display, but the oft-overlooked Laurentian Mountains paint just as pretty a picture. Carpeted in thick forest and scarred with icy blue lakes, Quebec City's local mountain range is vast and scarcely populated – even in its most captivating season.

As temperatures begin to drop to woolly jumper-wearing single digits and the air turns delightfully crisp, the trees here reveal a brilliant change of colour.

Contrasting against the intense blue skies, the landscape brims with vibrant red maple, vivid orange birch and golden-yellow larch. If this was New England, the paths would be busy with day-trippers, but in Canada, the forests – equally awash with colour – are hushed. The few leaf-peepers who venture here scatter across this gorgeous landscape: hiking to far-away viewpoints, soaring above the scenery on gondola rides or whizzing through the trees on ziplines (this is leaf-peeping at its most immersive). New England for fall? That's so last season.

Still going to New England?

If you want to see New England's fall colours without the crowds, venture off the road and up a mountain. The Percy Peaks Trail in New Hampshire offers epic views.

The kaleidoscope of colours on display in Quebec's Laurentian Mountains

Getting There *The Laurentian Mountains are just over an hour's drive north of Quebec City and two hours from Montreal.*

www.laurentides.com

Japan's cherry blossoms may draw attention in spring but nature is putting on a display all around the world. Put Japan on hold and seek the huge, blue hydrangeas in the Azores instead.

the alternative to *cherry blossoms in Japan*

HYDRANGEAS IN THE AZORES

Portugal

Spring's floral fiesta is nothing short of spectacular, and nowhere is that more true than during Japan's famous cherry-blossom season. If, though, you want to see a floral phenomenon without the crowds, head to the islands of the Azores. From April, their hills and valleys are smothered in hydrangeas, thousands of basketball-sized bursts of blue, purple and white that reach a crescendo of colour in July.

Hydrangeas are a symbol of the region and grow in abundance here, lining valleys in the lush interiors and dividing up the fields in lieu of fences. For the most spectacular displays, head to the isle of Faial. A volcanic eruption here in the 1950s made the soil more acidic and, consequently, the pigment in the petals more striking. Faial is often known as the "Blue Island", and if you visit at the end of summer, you'll see why.

Large hydrangeas bursting with colour on the "Blue Island"

Still going to Japan?

Head to a smaller city, like Okayama, or visit outside the main season – Shinjuku Gyoen in Tokyo is a good spot thanks to its early- and late-blooming trees.

Getting There *Faial is connected to Lisbon via its airport, Horta. Internal flights and ferries connect Faial with the other islands of the Azores.*

www.visitazores.com

MORE LIKE THIS

TULIPS
Turkey
The local government in Istanbul started planting tulips in 2006, and now the city is carpeted with millions of these beautiful yellow, white, red and purple flowers each spring.

WILD RHODODENDRONS
Nepal
Nepal's jagged, remote ambience is given a pink, photogenic burst each April as its colourful national flower bursts into bloom.

The rarely seen coral reefs that surround Christmas Island are just as splendid as those of the Great Barrier Reef in Australia – and come with plenty of bonus wildlife to enjoy on the side.

the alternative to the Great Barrier Reef, Australia

CHRISTMAS ISLAND

Australia

There's no denying that Australia's Great Barrier Reef is the most famous coral reef on the planet – due in no small part to its overwhelming size. But bigger doesn't always mean better, as you'll soon discover when exploring the incredible diversity of wildlife around Christmas Island.

Perched on the precipice of the Java Trench, this remote, rainforest-clad island rises majestically from the depths of the vast Indian Ocean. Astonishingly, it's home to more indigenous species than anywhere except the Galápagos Islands. The azure waters here form an aquatic Garden of Eden, brimming with more than 200 coral species – including plate coral over 3 m (10 ft) in diameter – and 600 species of tropical fish, not to mention a host of other creatures such as spinner dolphins, eagle rays and whale sharks. This is one of the few places you can see the spectacular but elusive dragon moray eel, as well as several rare endemic fish.

Your first descent into the crystal-clear water will be nothing short of breathtaking. Beneath the surface lies the most luxurious of hard coral reefs – a colourful, lush display that is almost ostentatious in its beauty. The view goes on forever, way down to the depths of the Indian Ocean. Brilliant rays of sunlight cast stripes of turquoise through the sea and highlight the flapping motion of a distant manta ray, as it approaches effortlessly and then slides on by. Inquisitive marble shrimp and playful butterfly fish dart between waving soft coral, while gentle wrasse and glum-looking groupers cruise calmly past the energetic scene. Listen carefully, and you might even hear the scraping symphony of rainbow-coloured parrotfish grazing on coral with their beak-like teeth.

A shoal of redfin, commonly found swimming among the corals of Christmas Island

Above *A diver entering Thundercliff Cave, which leads to a series of grottoes*

And in all that time there's no one else in sight other than your diving buddy and a divemaster. That fact alone is enough to give Christmas Island the edge over the ever-crowded Great Barrier Reef, but it's not the only thing that makes this place special. Millions of years of erosion have created countless underwater caves, providing a variety of diving opportunities beyond the reef. They reveal yet another world entirely: as you enter Thundercliff Cave, on the shore side of the fringing coral, the rippling rainbow of colour gives way to a series of pitch-black grottoes, decorated with monochrome spikes of stalactites and stalagmites. A little further on, you ascend to an extraordinary subterranean beach where you can climb out of the water and continue exploring on foot. It's not long before you realize you're inside an enormous cavern that's easily as beautiful and grand as any Gothic cathedral.

In fact, this is where Christmas Island really comes into its own – the natural wonders you'll find above the water are just as impressive as those encountered below. The whole island is a haven for birdlife, from the Abbott's booby, whose guttural call provides the soundtrack to the rainforest, to the Christmas Island frigatebird, whose males are instantly recognizable during mating season by their inflating red throat sacs.

Even more intriguingly, the island has been nicknamed the "kingdom of the crabs" for its extraordinary red crab migration: an annual spectacle that sees millions of the scarlet creatures emerge from the rainforest and march en masse down to the sea to spawn. They're just one of 20 crab species that can be found

Left *Red crabs migrating towards the ocean to spawn*
Below *Yellow forceps butterfly fish, one of the many hundreds of coral reef fish in the waters of Christmas Island*

here, along with football-sized robber crabs – the biggest land crustaceans on earth – so named for their propensity to collect shiny objects (including pots and cutlery from campsites and backpacks). This nature-lover's paradise really is the gift that keeps on giving – which seems only fitting for a place where every day it's Christmas.

Still going to the Great Barrier Reef?

Almost three million people visit the Great Barrier Reef each year. To experience this wonder in a small group, join a live-aboard boat tour to explore remote reefs and take night dives away from day-trip destinations. Time your visit to coincide with the annual mass coral spawning (November or December).

Getting There *Yacht moorings are available year-round; otherwise, flights to Christmas Island leave from Perth on mainland Australia.*

www.christmas.net.au

MORE LIKE THIS

GULF OF AQABA
Jordan

The city of Aqaba sits at the top of Jordan's small share of the Red Sea coast, but drive a short way towards the Saudi border and you'll find uncrowded coral reefs that run right up to the shore. They decorate the sand here like flower beds.

BANDA ISLANDS
Indonesia

The Banda Islands were long fought over for their nutmegs, once worth more than gold. This small brown nut is still the life-blood of the islands, but running a close second are its dazzling coral reefs, which have long been protected from fishers.

BLUE CORNER
Palau

In ancient times a matrilineal society regarded as one of the wealthiest in the Pacific, Palau is today a magnet for those few divers who are prepared to travel a long way for their sport – particularly for the famous Blue Corner dive site.

ART AND CULTURE

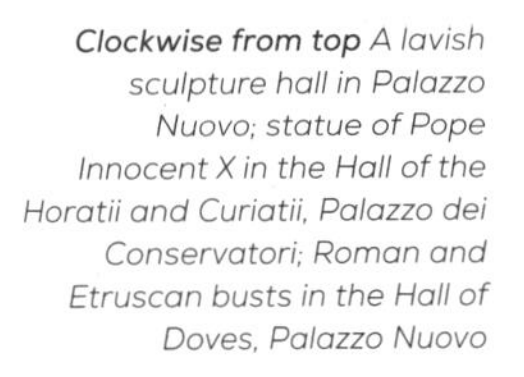

Clockwise from top *A lavish sculpture hall in Palazzo Nuovo; statue of Pope Innocent X in the Hall of the Horatii and Curiatii, Palazzo dei Conservatori; Roman and Etruscan busts in the Hall of Doves, Palazzo Nuovo*

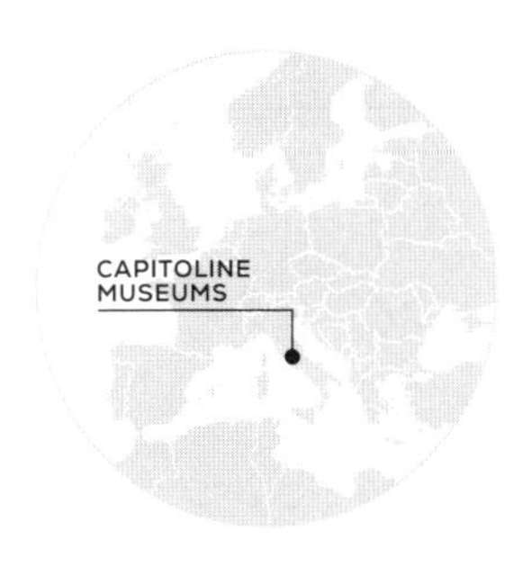

The Vatican Museums in Rome are the epitome of beauty, but with great beauty comes a great press of visitors – not so for the Capitoline Museums across the Tiber. Light and spacious, they make an elegant setting for some of the finest art of the antique world.

the alternative to *the Vatican Museums, Vatican City*

CAPITOLINE MUSEUMS

Italy

The Vatican Museums may have the Sistine Chapel, but you're likely to spend an hour queuing to see it – time better spent perusing the exceptional collections of classical sculpture and paintings at the Capitoline Museums. Lying on the gently rising Capitoline Hill, the Capitoline Museums stretch across the Palazzo Nuovo and Palazzo dei Conservatori. With no traffic and visitors never overwhelming (well, the Vatican Museums attract the lion's share of tourists – more than 4 million a year), it's relatively quiet here. The Vatican Museums across the river seem aloof and inhuman in scale by comparison.

There's plenty to rival the Sistine Chapel inside, too. Within the Palazzo Nuovo, an underground passageway leads to the Galleria Lapidaria, where the remains of the Temple of Veiovis lie. Passing into the courtyard of the Palazzo dei Conservatori, you're greeted by a jumble of fragments from a colossal statue of Emperor Constantine. Indeed, these museums are a divine exercise in both beauty and history. The story of Rome, for one, is told in a sequence of 16th- and 17th-century frescoes in the Hall of the Horatii and Curiatii. And the art collection keeps on giving in the picture gallery, where the vivid reds of Titian's *Baptism of Christ* and Garofalo's *Annunciation* are bound to leave you astounded.

Still set on the Vatican Museums?
Book a group tour *(www.vaticantour.com)* and you'll sail past the long lines. With a good guide, you'll also get a better handle on the collection in three hours than in days of dazed wandering on your own.

Getting There *The Colosseo metro stop is a 10-minute walk from the Capitoline Museums. Dozens of buses from Rome's main station, Termini, stop nearby too.*

www.museicapitolini.org

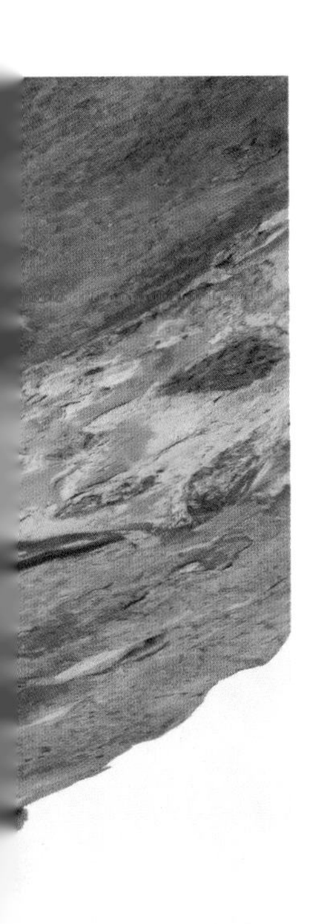

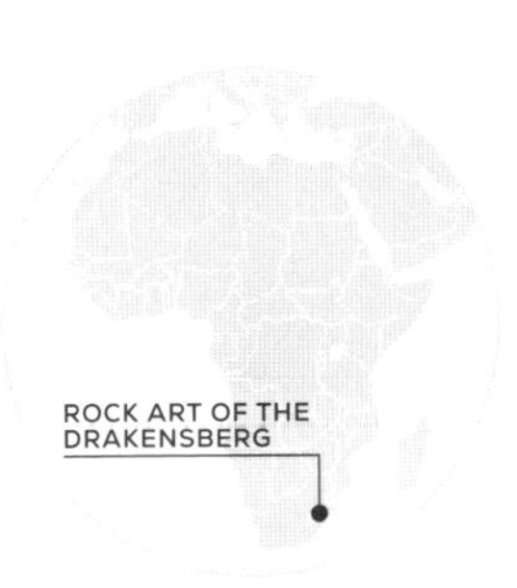

Lascaux may get all the glory when it comes to rock art, but due to conservation issues, visitors can only view copies of the prehistoric paintings here. To see something original, head to South Africa's Drakensberg mountains, where ancient art comes on a grand scale.

the alternative to
the cave paintings of Lascaux, France

ROCK ART OF THE DRAKENSBERG

South Africa

France's famous Lascaux Cave may be on your bucket list but, spoiler alert: it isn't the real thing. The original cave, which opened to the public in 1948 and featured painted figures by the Magdalenians, was closed just 15 years later due to damage caused by human traffic. What you see today, the Lascaux II cave, may be a faithful rendition of the original but for those looking to glimpse close up, and truly prehistoric rock art, it's time to book a ticket to South Africa.

Set in the Maloti-Drakensberg Park – Africa's largest tract of protected mountain wilderness – a sprawling open-air gallery comprises 500 painted shelters and caves. The ancient drawings here are remarkable for the fine quality of their craftsmanship, as well as their excellent state of preservation (they're all original, in comparision to Lascaux's copies).

The Game Pass Shelter, which features some of the area's most vivid rock art

Archaeologists estimate that the oldest paintings here are around 5,000 years old, while the most recent are perhaps just 150 years old – they depict the first ox-wagons to roll across the escarpment, carrying the European settlers who took possession of the land and killed the local population on sight.

The artists behind this extraordinary collection of work were the San people, who lived in this rocky area for around 4,000 years (and who now largely inhabit the Kalahari region). The San, members of various ethnic groups including the Khoe, Tuu and Kx'a, are believed to be descendants of the very first peoples of southern Africa. They drew panels in these caves using materials such as white clay, bird droppings, charcoal and blood, creating an earthy array of figures in red, white, brown and black.

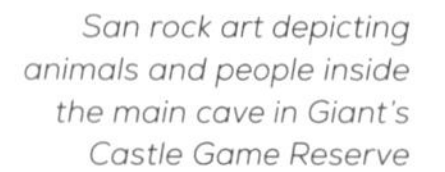

San rock art depicting animals and people inside the main cave in Giant's Castle Game Reserve

Above *Up-close details of one of the San rock art panels* **Right** *Trekking through the wilderness of the Drakensberg mountain range*

To visit any of the art in the park you need to book a rock art custodian, who will be your guide. For the best-preserved rock art in the area, head to the Game Pass Shelter in the Kamberg region. The panels here, showing animated figures in dance and various poses, were once thought to be a depiction of the San's daily life, but historian and archaeologist David Lewis-Williams (b 1934) believed they held more meaning. It was here that he "cracked the code" of the religious symbolism that underlies the area's rock art, revealing these paintings to depict figures in a state of shamanic trance and hunters absorbing the qualities of the animals they killed. Many now believe these caves were sacred spaces, a church if you like, and the paintings were an ancient form of stained-glass windows. The San didn't live in these caves, they only visited them during their most holy hours.

The Giant's Castle Game Reserve, in the centre of the park, also features a stunning display of rock art. The Main Cave here hosts 500 human and animal figures, including big cats, rhinos, snakes, baboons and the ubiquitous eland. The eland (Africa's largest antelope) was sacred to the San, for its spiritual value and physical sustenance.

Maloti-Drakensberg isn't just an incredible ancient art park. This vast area ranks among the elite group of 26 UNESCO World Heritage Sites to be inscribed on both cultural and natural grounds. Featuring spectacular scenery and abundant wildlife, its wild beauty is a world away from the landscaped walkways of Lascaux. Visitors can ramble through lush green foothills, brave a hike up to the escarpment or snake their way by jeep up the sole pass that breaches the otherwise impregnable border with

Lesotho. Far from being stage-managed, the Maloti-Drakensberg still hosts a profusion of the elands so beloved by the San, while quarrelsome baboons bark shrilly from the cliffs, vying to be heard above the calls of 300 species of birds. Remarkably, this backdrop, and its eclectic companny of inhabitants, has changed little since the San first started painting it.

Still going to Lascaux?

The "Sistine Chapel" of the prehistoric world is still one of the most important collections of cave paintings. A tour in Lascaux II can often feel crowded and rushed, so consider venturing to Lascaux IV instead. Here, you can see the reproduced panels as well as digital exhibitions explaining the art – which you can explore at your own pace.

Getting There *Hire a car at Durban or Johannesburg to drive to the Drakensberg (three or five hours away respectively).*

www.maloti-drakensberg.co.za

CUEVAS DEL POMIER
Dominican Republic
Just north of the city of San Cristobal are 55 caves containing the largest collection of rock art in the Caribbean. The paintings were created by the local Taínos and Igneris people and date back around 2,000 years.

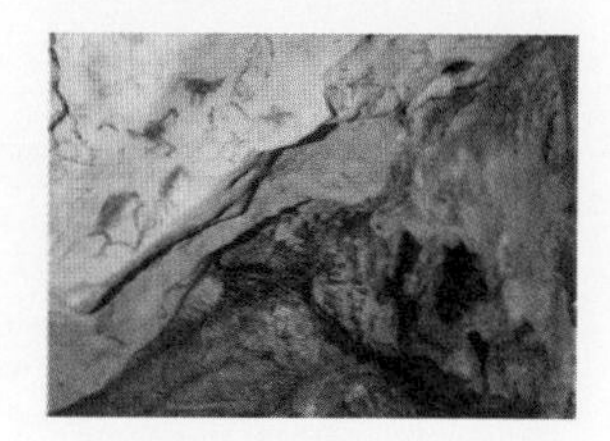

VALLÉE DES LES MERVEILLES
France
The Valley of Marvels in southern France holds the largest number of open-air petroglyphs in Europe. The drawings, which include cosmic symbols, are the work of the Mediterranean Bronze Age.

GOBUSTAN STATE RESERVE
Azerbaijan
This national park is the site of over 6,000 recorded carvings that depict people, animals, battle pieces, ritual dances and cosmology dating back 5,000 to 20,000 years. Mud volcanoes can also be found here.

Sure, Broadway gets all the fame, but Chicago's theatres could upstage a New York powerhouse any day. This Midwestern gem is the beloved birthplace of Kinky Boots *and the home of many a thought-provoking and genre-bending show.*

the alternative to *New York's Broadway, USA*

CHICAGO THEATRE SCENE

USA

The longest-running American musical on Broadway goes by the name *Chicago* – which seems like the perfect metaphor for the Windy City's dynamic theatre scene. While high-paying audiences gather in Manhattan, over in Illinois, Pulitzer Prize-winning writers like Tracy Letts and Bruce Norris are quietly creating ground-breaking plays. Want to see a work before it hits Broadway? Book a flight to O'Hare airport. Tony-award winning musicals like *Kinky Boots*, *The Producers* and *Spamalot*, as well as Lett's acclaimed play *August: Osage County*, are just a handful of the productions that premiered here.

The opulent interior of the Chicago Theatre, which is inspired by Parisian grand architecture

As in New York, the glittering marquees of the Theatre District in Chicago's Loop area have been drawing crowds since the 1920s. Celebrated destinations like the Goodman Theatre have garnered 22 Tony awards over the years, while the Chicago Theatre nearby has seating for 3,600 – nearly twice the size of the Gershwin Theatre, Broadway's biggest auditorium.

Theatrical entertainment here started much earlier than the 1920s, though. Most historians credit a certain Mr Bowers for giving Chicago's first performance. In 1834, the so-called Fire King drew a red-hot iron rod across his body – a mere prelude to his eating fire balls, before jumping into a second act of ventriloquism.

Fire played another key role in this city's history, when much of Chicago

State Street, with the bright sign of the 1920s Chicago Theatre, at the heart of the Chicago Theatre District

burned to the ground in 1871. Theatres were soon reborn, however, and today, as you stroll past the neon-lit façades and eager crowds near the corner of Randolph and State streets, you can feel the electricity in the air. It's easy to imagine that you're standing in the centre of the theatrical world. Broadway – isn't that a street over on the north side, near Wrigley Field?

Still want to see a show on Broadway? Head to the TKTS Booth in Times Square, where you can get same-day discounted tickets for both top Broadway and off-Broadway shows.

Getting There *O'Hare International Airport is Chicago's main transport hub. Head to the Loop (downtown) to see a show.*

www.chicagoplays.com

TBILISI OPERA AND BALLET THEATRE

Want to see ballerinas from the Bolshoi and other equally renowned companies without the price tag? Tbilisi Opera and Ballet Theatre has you covered.

the alternative to
the Bolshoi Theatre, Russia

TBILISI OPERA AND BALLET THEATRE

Georgia

Founded in 1851, Tbilisi Opera and Ballet Theatre occupies a stunning neo-Moorish building in the centre of Georgia's capital. But despite the glitzy interiors, this is by no means an exclusive affair: tickets are priced so everyone from students to families can indulge their passion for culture. The highlight of this venue's annual calendar is the Tbilisi Ballet Festival. Some of the biggest names – including the principal dancers of the Bolshoi and Joffrey ballets – take to the stage here. So, to see a dazzling variety of ballet in a gorgeous venue, make for Tbilisi instead of Moscow.

Still going to the Bolshoi? If ballet tickets appear to be sold out, visit the box office to check for returns.

Getting There *Fly into Tbilisi International Airport. From here it's best to hop on a bus or take a taxi to the city centre.*

www.opera.ge

A work of art in itself, the fittingly futuristic New Museum in New York City outshines the more famous Museum of Modern Art as a showcase for truly cutting-edge contemporary work.

the alternative to
the Museum of Modern Art, USA

THE NEW MUSEUM

USA

A stack of aluminium boxes, shimmering 53 m (175 ft) above Manhattan's gritty Bowery, the New Museum looks like an alien superstructure looming over New York's Lower East Side. Designed by Japanese architects Kazuyo Sejima and Ryue Nishizawa and opened in 2007, it was joined in 2022 by an equally futuristic Rem Koolhaas-designed addition, with a glass and a mesh-like façade. It's as though a piece of the *Star Wars* set has been marooned amid the tenement buildings.

While the city's Museum of Modern Art (MoMA) is vast, sprawling and a little formal, the New Museum is its smaller, more laid-back cousin. Here, the art is often shocking and the spaces no-frills – the building's guts are exposed on every floor. And in contrast to New York's more established art museums, the exhibition rooms at the New Museum are huge. An industrial elevator glides between the four main floors, each holding a single, high-ceilinged gallery. These warehouse-style rooms are luminous spaces, all brilliant white with polished concrete floors.

The New Museum's diverse range of changing exhibits are different, too – a little more edgy, a little more challenging than MoMA's more conventional roster. Shows, which feature both emerging and established contemporary artists, have included a retrospective of Harlem-born artist and activist Faith Ringgold, Lynn Hershman Leeson's disturbing *Breathing Machine* sculptures (ghoulish

Visitors browsing works of art in one of the New Museum's light-filled galleries

The striking exterior of the New Museum, the only museum of contemporary art in Manhattan

wax portraits with wigs and sound) and Daiga Grantina's ode to the dynamic properties of lichen, the vividly named *What Eats Around Itself*.

Ingeniously, the building itself even actively enhances the viewing experience: the gallery interiors have been designed to utilize natural daylight (through sky-lights), which means the art seems to present subtle changes at different times of the day, and in different spaces. At this museum you really are always guaranteed to see something new.

Still going to MoMA?

New York's Museum of Modern Art is one of the world's most popular storehouses of art, and is often very busy, with long lines to get inside. It's often best to visit towards the end of the day, during the week, when the galleries start to empty – mornings are usually much busier. Try to avoid weekends altogether.

Getting There *The New Museum is close to the Second Avenue subway station (F train), as well as Bowery station (J, Z).*

www.newmuseum.org

MORE LIKE THIS

INTERNATIONAL CENTER OF PHOTOGRAPHY MUSEUM
USA

All things photography are celebrated at this sleek, dynamic museum, which moved into new digs at Essex Crossing in New York's Lower East Side in 2020.

THE NOGUCHI MUSEUM
USA

Devoted to Japanese-American abstract sculptor Isamu Noguchi (1904–88), this museum in Queens, New York, displays the world's most extensive collection of the artist's work. It pairs well with a visit to the nearby Socrates Sculpture Park.

MOMA PS1
USA

Most visitors skip the more avant-garde Queens outpost of MoMA (or aren't even aware that it exists in the first place), but it's well worth a visit for its range of fascinating and often challenging exhibitions, across a variety of media.

Tate Modern's towering industrial setting may be the go-to for the contemporary art scene but it's not unique. Zeitz's converted grainstore venue hosts just as exciting exhibitions, with the added focus on ground-breaking African art.

the alternative to *Tate Modern, UK*

ZEITZ MUSEUM OF CONTEMPORARY ART AFRICA

South Africa

The tubes of the former grain silo, fixed with huge pillowed glass windows, standing tall in Cape Town

Tate Modern might be hard to beat when it comes to international art, but the Zeitz Museum of Contemporary Art Africa (MOCAA) in Cape Town isn't trying to compete. Rather, it's the specificity of this collection that makes a visit compelling, showcasing the contemporary culture of South Africa, the African continent and the African diaspora. There's no better way to immerse yourself in African culture than visiting an art house by Africans, in Africa.

Housed in a building worthy of its own exhibition, the Zeitz MOCAA's stunning space was carved out from the concrete tubes of a grain silo built in 1921. Technically, this nine-storey gem, planned by British designer Thomas Heatherwick and comprising 100 individual galleries, is an even bigger feat than the transformation of an old London power station into Tate Modern. Inside, white cube-like gallery spaces radiate from the cathedral-like atrium at the heart of the museum, a sculpture garden sits on the rooftop, and bookshops, an art conservation facility, restaurant and even a luxury hotel fill the epic space that remains.

But it's the art that shines here, with artists using the gallery spaces to delight, question, defy and confirm African identities. Take Anthony Bumhira, who uses doilies in his large-scale abstract works. These crocheted pieces are an intricate part of life in Zimbabwe, and Bumhira tells the story of mothers who spend hours creating and selling these decorative and functional table pieces in order to feed their families. Or photographer Zanele Muholi, whose bold and fierce work depicts the lives

Below *Unusual atrium inside the museum, made up of the silos' excavated innards*
Bottom *Curator Tandazani Dhlakama explaining the intricacies of a painting*

of the continent's LGBTQ+ community, both confronting prejudices and celebrating "otherness".

It's an understatement to say the works here are inspiring. Banele Khoza's pastel male nudes, for example, may have earned him criticism in his native eSwatini, but in South Africa his work is exhilarating. That's the way with the continent: it speaks many languages in many ways, and the artistic storytellers displayed at the Zeitz MOCAA know every word.

Still going to Tate Modern?

With over 70,000 works of modern art, the Tate Modern always makes for a thought-provoking visit. Swerve the school groups crowding round Picasso and Rothko and seek out the less mainstream artists on display.

Getting There *The gallery is easily accessible from Cape Town city centre whether you're using your hotel's transfer facilities or taking a bus or taxi.*

www.zeitzmocaa.museum

Housing within its grand walls some three million specimens up to 570 million years old, Argentina's intriguing Museo de La Plata is a fine competitor of Washington's Smithsonian.

the alternative to
Smithsonian Natural History Museum, USA

MUSEO DE LA PLATA

Argentina

Founded in 1884 to exhibit the discoveries of explorer Francisco P. Moreno, the Museo de La Plata assembles the remarkable history of South America under one roof. While a trip to Washington's museum (the world's most visited natural history museum) can be a whirlwind – as you zip past ogling crowds ticking off the big hitters – Argentina's lesser-visited museum offers a fascinating time-travelling adventure.

Inside these hushed halls, you'll find a replica of a diplodocus skeleton plus the fossilized skin of a Mylodon – a giant sloth-like creature that roamed Patagonia some 10,000 years ago. The ethnographic and archaeological rooms are also a treat. From pre-Colombian ceramics to audio-visual displays, the museum expertly introduces visitors to the cultures of those who came before us.

One of the museum's most popular exhibits, the 27-m- (89-ft-) long copy of a diplodocus skeleton

Still going to the Smithsonian?
For the quietest experience, visit on a Monday, Tuesday or Wednesday, when museum crowds are at their minimum.

Getting There *The museum is in the city of La Plata, an hour's drive from Buenos Aires. Trains and buses also connect the two cities.*

www.museo.fcnym.unlp.edu.ar

MORE LIKE THIS

TE PAPA
New Zealand
As well as exploring Māori history, this vast museum features contemporary art and an outdoor nature exhibition – complete with a bushwalk.

SHANGHAI NATURAL HISTORY MUSEUM
China
With its curving, glassy exterior, this architecturally acclaimed museum houses everything from dinosaur skeletons to ancient Chinese pottery.

A bite-sized version of the Prado (which houses the most comprehensive collection of European paintings in the world), the Palacio de Liria provides a dazzling overview of European art that's much easier to digest.

***the alternative to** Museo del Prado, Spain*

PALACIO DE LIRIA

Spain

The Prado holds so many works of art that if you visited the museum all day, every day, it would take around six months to see them all. To feast your eyes on some equally fine art, but without feeling overwhelmed, head across Madrid to the Palacio de Liria. The Duke of Alba's exquisite mansion was opened to the public for the first time in 2019 and is home to one of the world's finest private art collections.

Like the Prado, this palace is an 18th-century Neo-Classical building. It was designed by Spanish architect Ventura Rodríguez and modified by the English architect Sir Edwin Lutyens, before being rebuilt after its interior was gutted by fire during the Spanish Civil War. The rooms were redone lavishly, and each is crammed with paintings, sculptures, tapestries, furniture and books – an eclectic collection that has been amassed over half a millennium by the Dukes of Berwick and Alba.

Most of the rooms are themed, with the Salón Flamenco, the Flemish Room, showcasing work by Peter Paul Rubens and Jan Brueghel the Elder. The Salón Español is the setting for Goya's striking *White Duchess* portrait – a highlight of the Spanish master's repertoire.

The Salón Italiano, with many notable Italian Renaissance paintings, including works by Perugino and Titian

Still going to the Prado?

Come in the evening (6–8pm Monday to Saturday, 5–7pm Sunday), when the museum is quieter and entry is free.

***Getting There** The nearest airport is Madrid-Barajas Airport. The museum is right next to Ventura Rodríguez metro station.*

www.palaciodeliria.com

Ballet and opera were born in Italy, but Brazil has lovingly cultivated these art forms for centuries. Besides, when you can say you've watched opera in the heart of the ambrosial rainforest, why wouldn't you choose Teatro Amazonas over La Scala?

The Venetian chandeliers illuminating the painted ceiling and curved balconies of the main auditorium

the alternative to *to La Scala, Italy*

TEATRO AMAZONAS

Brazil

Every corner of Manaus's palatial Amazon Theatre, behind its soft pink façade punctuated by crisp white trimming, is opulence defined. It was modelled on La Scala, after all, so you get the same level of beauty as Milan's celebrated theatre but with an added bonus: you're in the Amazon. It's a given to find a theatre in a city, but in the jungle? La Scala simply cannot compete when it comes to setting.

Construction began in 1884, in the wake of an economic boom from rubber plantations and the free labour of the enslaved in the city of Manaus. Not a single *centavo* was spared on the elaborate design: the roof's red tiles were imported from Alsace, the columns and stairs chiselled from Carrara marble, the chandeliers brought from Venice to warmly light the celestial scenes of music, dance and drama painted on the ceiling of the auditorium. The details are not all European, though. The stage curtain painting, *Meeting of the Waters*, portrays the nearby natural wonder Encontro das Águas, where the dark waters of the Black River meet, but do not mix with, the cafe-au-lait waters of the Solimões River. A beautiful touch, this artwork is another delightful reminder that you're in the heart of the Amazon.

From 1897 to 1911, Italian, Portuguese and French companies took to this fine stage, entertaining the wealthy who patronized the grand theatre. However, 14 years after the first performance, the theatre closed for a 90-year hiatus. The invention of artificial rubber, coupled with the abolition of slavery in 1888, sapped Manaus of its riches, while World War I put a stop to European companies visiting. The building was nonetheless maintained, and in 2001, the provincial government reopened the theatre after major renovations. It's flourished ever since.

Musicians and performers from around the world have made their creative homes here, producing a rich programme. The

Below An orchestral performance in the theatre
Bottom The splendid pink-and-white façade of Teatro Amazonas

Amazon Philharmonic Orchestra graces the stage during the annual opera festival from April to May, while dance festivals, concerts from international bands and local artists, and the Amazonas film festival run on occasions throughout the year. From the beginning, the theatre's goal was to provide an elite arts experience, and to do so in high fashion. Consider it achieved.

Still seeing a show at La Scala?
This 18th-century theatre is dubbed the world's greatest opera house for good reason – Verdi was once house composer, and costumes are designed by top fashion names. The season runs from December to May, and its vital to book well in advance to secure tickets. Word to the wise: the foyer bars are often crammed before a show, so have a pre-drink at Leonardo's Vineyard *(La Vigna di Leonardo)* nearby instead.

Getting There *Manaus International Airport (MAO) is a 30-minute drive via car, taxi or airport shuttle to the Amazon Theatre.*

www.cultura.am.gov.br/portal/teatro-amazonas

Parisians are happy for visitors to flock to the Louvre, as that way they keep the Musée Jacquemart André all to themselves. This splendid former home is a time capsule filled with lovingly curated objets d'art.

the alternative to the Louvre, France

MUSÉE JACQUEMART ANDRÉ

France

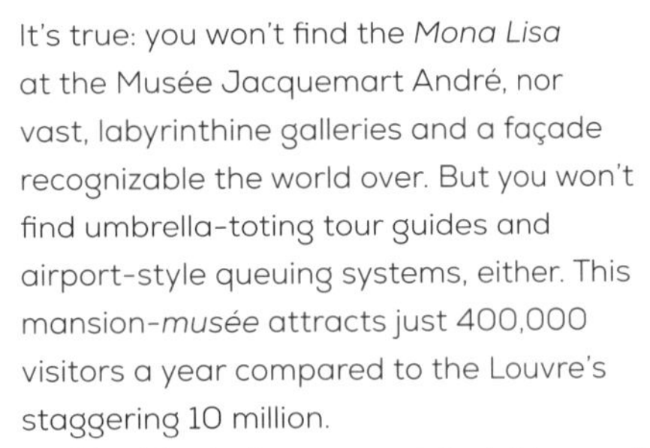

It's true: you won't find the *Mona Lisa* at the Musée Jacquemart André, nor vast, labyrinthine galleries and a façade recognizable the world over. But you won't find umbrella-toting tour guides and airport-style queuing systems, either. This mansion-*musée* attracts just 400,000 visitors a year compared to the Louvre's staggering 10 million.

The Musée Jacquemart André is proof that good things come in small packages. (Well, relatively small at least.) Beyond the Second-Empire splendour of the museum itself – a palatial pile just off Boulevard Haussmann, between Parc Monceau and the Champs-Élysées – there's plenty to marvel at. The drawing rooms and state apartments overflow with paintings, tapestries and sculptures by some of Europe's greatest artists: Rembrandt, Botticelli, Canaletto and Van Dyck, to name just a few.

The collection here has a personal feel, not least because the building was originally the private home of banking heir Edouard André and his wife, painter Nélie Jacquemart. Nélie had a passion for the Italian Renaissance, and the walls are adorned with works the pair acquired while on their travels. Many rooms are preserved as they designed them, including the Picture Gallery, where the same pieces greet guests today as they did at the couple's lavish parties in the late 1800s.

The glass-roofed Winter Garden (Jardin d'Hiver), adorned with plants, a grand staircase and marble paving

The Picture Gallery (Salon des Peintures), filled with splendid paintings and furnishings

Save the private apartments for last, for a touching glimpse of the portrait of Edouard that Nélie painted in 1872 – it was through this commission that the pair first met. These sorts of details make this much more than just a museum: it's a window into a life and whole other world.

Still going to the Louvre?
There are several tricks to getting the most out of the Louvre. Buy tickets for timed entry and map out a few pieces or collections you want to see before you get disoriented. Audio guides are well worth the small fee.

***Getting There** The nearest Metro stations are Saint-Augustin, Miromesnil and Saint-Philippe du Roule, via lines 9 or 13.*

www.musee-jacquemart-andre.com

Nepal's Patan Museum may be less lauded than the National Museum in Delhi, but it punches well above its weight in cultural terms, thanks to a breathtaking array of artifacts.

***the alternative to** the National Museum, India*

PATAN MUSEUM

Nepal

Few institutions can compete with the National Museum in Delhi's compendious array of artworks, but who's to say you need thousands of exhibits to provoke a sense of awe and inspiration? Occupying the northern wing of a 17th-century palace in Patan (a city bordering Kathmandu), on a square lined with historic temples, the Patan Museum is worth a visit for the architecture alone. Inside, its extensive collection ranges from sacred Buddhist and Hindu iconography – including many intricate woodcarvings, bronzes and sculptures – to the gold-encrusted thrones of the Malla kings who ruled Nepal for centuries. Together, they offer an illuminating insight into Nepal's diverse heritage – sometimes less really can be more.

Still going to the National Museum?
Home to around 200,000 exhibits, Delhi's National Museum can be overwhelming. Take the "Museum in 90 minutes" tour for an introduction to the collection's highlights.

***Getting There** Patan has good bus connections to neighbouring Kathmandu and the surrounding Kathmandu Valley.*

www.patanmuseum.gov.np

Amsterdam's Van Gogh Museum may have the world's largest collection of the painter's work, but it's about quality, not quantity for the Kröller-Müller. For a taste of Van Gogh in a more tranquil setting, leave the city behind and head into the Dutch countryside.

Admiring Van Gogh's famed Country Road in Provence by Night, *also known as* Road with Cypress and Star

the alternative to
the Van Gogh Museum, The Netherlands

KRÖLLER-MÜLLER MUSEUM

The Netherlands

Art-lovers anxious to get their fix of Van Gogh should bypass the popular Van Gogh Museum in Amsterdam, with its overcrowded rooms and long queues, and head instead to the fascinating Kröller-Müller Museum. Here you'll find an extensive collection of the great man's work in an idyllic rural setting of woodland, heathland, grassy plains and sand drifts – sublime beauty akin to that of a Van Gogh painting.

This fine, isolated museum is named after Helene Kröller-Müller, a 20th-century art enthusiast who built up a large collection of works, including many Van Gogh paintings, with the support of her wealthy industrialist husband. But when his business was hit hard during the Great Depression, Helene feared for the future of her collection, and in 1935 she gave all 11,500 pieces to the Dutch state on the condition that a museum be built to house them. Her wish was granted, and in 1938 the Kröller-Müller Museum opened on the family estate, located in the heart of Hoge Veluwe National Park in the eastern Netherlands.

Helene's favourite artist was Van Gogh, who she believed "created modern Expressionism". She purchased 91 of his paintings and 175 drawings, and, while this is less than half the number owned by the Van Gogh Museum in Amsterdam, it represents a more digestible helping for one visit. Look out for the arresting *Self Portrait* of 1887, in which the artist's face emerges from a thunderous, swirling background and seems to stare right at you, and *Four Cut Sunflowers*, with its fiery yellow petals that appear to flicker on the canvas. The paint has been applied so thickly in *The Sower* (1888) and *Country Road in Provence by Night* (1890) that the works have a shimmering quality, the energy from the bright stars and the sun somehow radiating far beyond the edges

of the canvases. The famous piece *Café Terrace at Night* (1888) is also on display; some critics believe this is part of a trilogy that includes *Starry Night* (1889) and *Starry Night Over the Rhone* (1888), both of which are exhibited elsewhere.

Van Gogh isn't the only reason to visit the Kröller-Müller, which can't be said for the Van Gogh Museum. Thanks to Helene's impressive art collection, you'll also get to feast your eyes upon a major display of 19th- and 20th-century French paintings and a sculpture garden with distinctive works by Auguste Rodin and Henry Moore. Now that's bang for your art buck.

Still set on the Van Gogh Museum? Visit at either end of the day to avoid the worst of the crowds. Once you're in, don't try to see too much: stick to one floor and take your time soaking up the works.

Getting There *Schiphol Airport is 95 km (60 miles) away from the Kröller-Müller. You can travel to the museum by car, train or bus.*

www.krollermuller.nl

Top Café Terrace at Night, *one of Van Gogh's most famous works in the museum* ***Above*** Floating Sculpture *(1960) by Marta Pan in the sculpture garden*

CAPTIVATING CITIES

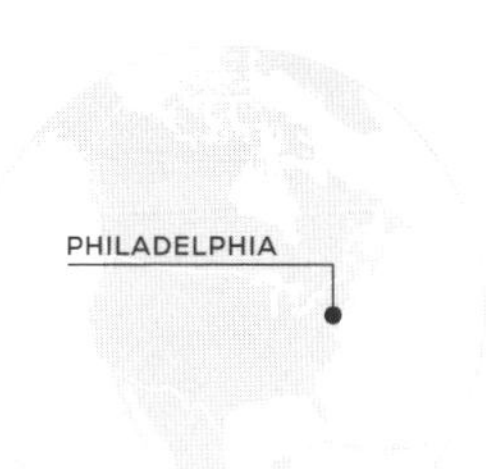

With its array of historical sights, bold dining scene and kaleidoscopic street art, Philadelphia is more than a match for iconic New York City. This is a city fuelled by an independent spirit, and all the better for it.

the alternative to *New York City, USA*

PHILADELPHIA

USA

New York has it all: soaring skyscrapers, an unimaginable number of restaurants and some of the world's greatest galleries. Few would think that the unassuming, and lesser-visited, Philly could give this iconic city a run for its money. Think again.

Founded by the Quakers in the late 17th century, the City of Brotherly Love straddles two very different worlds. As one of America's oldest cities, Philadelphia wears its past with pride, showcasing key monuments in a UNESCO-listed historic district. And yet, like New York, Philly also has a dense urban core that's home to a treasure trove of foodie and arty delights.

Yes, Manhattan has its share of historic buildings, but you won't find the same concentration of venerable sites as you will in Philadelphia's Old City, sometimes referred to as America's most historic square mile. Here, visitors can stroll the corridors of Independence Hall, where early American visionaries hatched the first seeds of an independent nation; you can also participate in the excitement, via multimedia exhibits at the Museum of the American Revolution. Well-preserved 18th-century mansions can be found throughout the city, like the Hill-Physick House, the former home of Philip Syng Physick (the so-called father of American surgery). If eyeing vintage stomach pumps leaves you feeling a little woozy, head to the nearby City Tavern to steady yourself over a historic libation. Try the Poor Richard's Tavern Spruce, a beer based on a recipe devised by Benjamin Franklin.

Befitting a city of its size, Philadelphia is home to some outstanding museums. The Philadelphia Museum of Art houses a world-class collection of art, as well as a ceremonial teahouse from Japan.

Broad Street, with the 19th-century Philadelphia City Hall and more modern architecture

***Above** An array of restaurants in Philly's bustling Chinatown district*
***Right** Mosaic artworks in Philadelphia's Magic Gardens*

***Right** The Benjamin Franklin Parkway and Downtown seen from the Philadelphia Museum of Art*

There is even a complete 18th-century drawing room from London's Lansdowne House here, too. The museum features another iconic site: the "Rocky Steps". In the 1976 film *Rocky*, Sylvester Stallone charged up the 72 steps to this museum's entrance and raised his arms in glory while gazing at the skyline – you'll likely see fans doing the very same.

New York has an impressive collection of street art, but even the colour-splashed stretch of Bushwick can't compare to Philly's mass of outdoor art. Founded in 1984, Mural Arts Philadelphia has added nearly 4,000 large-scale artworks to the public space, and the city can rightfully lay claim to being the world's largest outdoor art gallery. Another outdoor art highlight is Philadelphia's Magic Gardens. The artist Isaiah Zagar spent 30 years transforming once empty city lots into a fantastical folk-art confection of mosaics and murals made from repurposed mirrors and ceramics as well as found objects like bicycle wheels and dinner plates. Strolling through this shimmering labyrinth is like entering another realm, one where you half expect tiny pixies to be lurking around the next tile-trimmed corner.

You'll find more Michelin-starred restaurants in New York, but Philadelphian locals will tell it to you straight: wherever you roam here, expect to pay less and eat better than you would in New York City. Teeming food halls like the Reading Terminal Market showcase the city's culinary diversity, with dozens of stalls offering everything from steaming Pennsylvania Dutch-style dumplings to spicy Thai curries. The open-air purveyors of the South 9th Street Italian Market serve traditional delicacies (plus tacos

and chilaquiles from more recent Philly immigrants), while Chinatown is packed with an astonishing variety of flavours. Visit imaginative fusion spots like Cheu that blend Japanese, Korean and even Jewish recipes in deliciously daring mash-ups.

Despite its wide-ranging appeal, Philadelphia doesn't trumpet its grandeur. Locals are proud of their rough-around-the-edges reputation. Like Rocky, they relish being the underdog; they wouldn't have it any other way.

Still want to visit New York City?

You can avoid high Manhattan prices by staying just across the East River. Long Island City has dozens of hotels, and it's a short subway ride to Midtown.

Getting There *Philadelphia International Airport has flights across the USA and links to international destinations. There's also good rail and bus service.*

www.visitphilly.com

BALTIMORE
USA

Charm City lives up to its moniker with an atmospheric harbour, old-school diners and one-of-a-kind sights like the American Visionary Art Museum, which celebrates outsider artists. Visit in the summer when festivals are in full swing.

JOHANNESBURG
South Africa

South Africa's biggest city has a lot to offer: skyscrapers hover above the city's business centre while cool areas such as Maboneng promise rooftop bars and quirky galleries. The Apartheid Museum is also a must-visit here.

MONTERREY
Mexico

Founded in the 16th century, Monterrey is a vibrant, and often overlooked, city in northern Mexico. Its historic district is packed with bars and restaurants, while serene green spaces pepper the centre and mountains loom in the distance.

Progressive and creative, Leipzig is a lot like Berlin. Saxony's coolest city may be smaller than Germany's capital but with its pulsing sense of possibility and affordable rents, it remains a magnet for artists, musicians and entrepreneurs.

the alternative to Berlin, Germany

LEIPZIG

Germany

True, there are many similarities between Leipzig and Berlin. Both have happening music and nightlife scenes, thriving subcultures, and oodles of galleries, bars, cafés and restaurants that open (and close) with regularity. But while Berlin's allure remains firmly intact, and rightly so, rising rents and gentrification have made it more difficult for up-coming creatives to find their place. Leipzig, on the other hand, is like the Berlin of 30 years ago, with enough uncut edges and underground venues to make you feel like there's still something left to discover or create.

And that's exactly what artists, musicians and entrepreneurs have been doing here since the early 2000s, when people were drawn back to this city a decade or so after the fall of the Berlin Wall. Students flock to the HGB (Academy of Fine Arts Leipzig), the former East Germany's most important art school, whose galleries exhibit their exciting new work. They also come to the University of Leipzig, which has an impressive list of graduates that includes playwright Johann Wolfgang von Goethe and the former German Chancellor Angela Merkel (yes, she grew up in East Germany).

Evidence of this young, make-it-on-your-own populace can be seen all over the city in small galleries, exhibition spaces and independent stores. At the epicentre of this scene is the much-loved Baumwollspinnerei, an old cotton mill and now an extensive arts complex of galleries and studios (including that of German avant-garde painter and HGB graduate Neo Rauch). This place has endless displays of experimental art and might have you stopping in at the arts supply store before you leave.

Not into fine art? The city's extensive events calendar rivals that of Berlin's in diversity and scope. The DOK Leipzig (a documentary and animated film festival) and the vast Christmas Market entertain in winter. Summer sees the Wave Gothic Festival (the world's largest celebration of goth culture) and the famous Bachfest. There's much more besides all year round.

The mix of contemporary and 19th-century architecture that makes up Leipzig's cityscape

Top A busy street café in the centre of the city *Above* The Karl-Heine Canal in the trendy Plagwtiz district, lined with brick warehouses converted into artists' studios

There's talk that a few of Leipzig's neighbourhoods, like industrial-turned-hip Plagwitz for example, have started to see some gentrification. But there are still plenty of old-school taverns with a whiff of the former East Germany that make you feel like you've travelled back in time. If you're not up for a vegetarian *Currywurst* or vegan Döner kebab, don't worry. You can still kick back in a *Kneipe* (pub) serving up *Fassbier* (draft beer) and classic German comfort food. Just don't wait too long to get here – this is Germany's fastest-growing city and word is out.

Still heading to Berlin?

Berlin is flat and its bike lanes make cycling an easy way to get around if you want to avoid public transport. There are several bike-sharing schemes.

Getting There *Frequent trains from Berlin reach Leipzig in less than an hour-and-a-half.*

www.leipzig.travel

With many of the best things you'll find in Marrakech, Taroudant is a scaled-down version of the famous larger city. Relax in this little-known gem, which exudes an easy-going charm.

***the alternative to** Marrakech, Morocco*

TAROUDANT

Morocco

They call Taroudant "the mini Marrakech", for a reason: this small town near Agadir has bite-size pieces of the best things that Marrakech offers. It may not have the gourmet dining, upmarket riads or the nightly open-air circus of Jemaa el Fna (Marrakech's main square), but Taroudant provides a real snapshot of Moroccan life.

Briefly the capital of Morocco in the 16th century, Taroudant preserves an air of grandeur. Unlike sprawling Marrakech, the whole town is elegantly surrounded by the city's perfectly preserved golden-brown walls. Sunset, when the stone is bathed in a warm glow, is the best time to walk or cycle around their perimeter and admire the imposing gates. Sundown is also when the otherwise sleepy main square, Place al Alaouyine, comes to life, with townsfolk gathering in the cool evening air. You might catch the odd storyteller or musician at weekends.

Leading off the square, the maze of little alleyways is a pleasure to get lost in. Don't miss the two daily souks, which are a gift to all the senses. Here you can indulge in some low-stress retail therapy for local crafts from the nearby Anti-Atlas villages and shop for fruits, vegetables and spices. There's loads to choose from, but haggling is a laid-back affair. Kick back with a cup of traditional mint tea offered by the vendors and soak it all in.

Still going to Marrakech?

Stay in an authentic riad where you can enjoy a calm oasis with a central courtyard and fountain. Comfortable, inexpensive options abound.

***Getting There** Agadir Al-Massira International Airport is 65 km (40 miles) from Taroudant. The easiest way to reach Taroudant is by shared taxi.*

www.visitmorocco.com/en/travel/taroudant

One of the five entrance gates to the 16th-century medina (old city) of Taroudant

Ghent matches Amsterdam's picturesque canal-side location and artsy creativity but has the kind of relaxed historic centre that has become a distant memory for the Dutch city.

Step gabled houses lining the Graslei quay in the old centre of Ghent

***the alternative to** Amsterdam, The Netherlands*

GHENT

Belgium

Amsterdam is a special place, of course. World-class art museums, waterside cafés, a creative vibe. But take a look just over the border with Belgium and you'll find another beautiful old city cut through with canals. Unlike its Dutch double, Ghent is a place to absorb slowly – on foot rather than on two wheels – with a pedestrianized centre that's home to plenty of architectural showstoppers. Spot Gravensteen Castle, the Stadhuis and the Graslei, a pastel-coloured row of step-gabled houses reflected in the waters of the Leie.

If you think historical Ghent can't rival Amsterdam for contemporary arts, think again. Werregarenstraatje is a canvas for street artists, Gentse Feesten is a vibrant music and theatre festival and, thanks to the large student population, there's a glut of vegan restaurants and trendy bars. Ghent does cool cafés too – and the strongest thing you'll find on the menu is a finely blended Arabica.

Still going to Amsterdam?

Avoid weekends (Fri–Sat), book tickets in advance and visit museums during their extended opening hours in the evenings.

***Getting There** Ghent is 65 km (40 miles) from Brussels International Airport. Trains from Brussels Central reach Ghent in an hour.*

www.visit.gent.be

MORE LIKE THIS

AMERSFOORT
The Netherlands
This canal-laced city, 50 km (30 miles) southeast of Amsterdam, has a museum dedicated to the Dutch abstract artist Piet Mondrian.

ZIERIKZEE
The Netherlands
Laden with historical monuments, this port city on the North Sea coast has a compact centre dating to the Middle Ages.

A city of freethinkers in America's Northwest, Portland proudly touts its indie credibility – which rivals San Francisco's in everything from dining and drinking to eco-friendly design.

the alternative to *San Francisco, USA*

PORTLAND

USA

Like San Francisco, Portland has a deep-rooted countercultural spirit. You can see it in bumper stickers and signs around town, proclaiming "Keep Portland Weird". Indeed, it doesn't take long to uncover Portland's bizarre side – a little time pondering vintage machinery at Stark's Vacuum Museum or looking for leprechauns at the tyre-sized Mill Ends Park (the world's smallest park) will soon clue you in.

This independent ethos infuses all aspects of life in the city, particularly when it comes to food and drink. A commitment to local and sustainable ingredients is just as important here as it is in San Francisco, and in terms of originality, Portland's young chefs tend to push the boundaries more than their Californian counterparts. You'll find restaurants offering creative mash-ups like Brazilian-Japanese fusion, menus filled with unusual wild and foraged ingredients, and dishes that elevate the cooking of vegetables to high art.

Even the food trucks (or "food carts", as they're known in these parts) are on a whole other level – with more than 500 scattered around the city, there are nearly three times as many as in San Francisco. The breadth of variety is staggering: you can munch your way around the globe, from Salvadoran *pupusas* to Vietnamese *bahn mi*. And the choice of drinks to wash it down with is no less vast – Portland is one of America's beer capitals, with over 70 microbreweries. (It's also got a legendary coffee scene for good measure.)

Daring to be different just seems to come naturally here. San Francisco may have the Golden Gate Bridge, but Portland (sometimes called "Bridge City" for its dozen unique spans over the Willamette River) is home to the much-loved Tilikum Crossing. This became

The cable-stayed Tilikum Crossing (also known as The People Bridge), dramatically illuminated at night

America's largest car-free bridge when it opened in 2015, and has been a magnet for cyclists and strollers ever since. Even more wonderfully, the entire structure becomes a living art installation by night, with 178 LED lights changing colour and pattern based on the river's water temperature and rate of flow.

If this is what weird looks like, then no wonder the locals are so passionate about keeping it that way.

Still going to San Francisco?

Don't bother renting a car. Aside from the challenges of driving (steep hills, narrow streets, pricey parking lots), San Francisco has neighbourhoods ideal for exploring on foot and decent public transport. If you come in the summer, prepare for chilly weather as cool fog can blanket the city.

__Getting There__ Portland has its own train station and international airport. It's about a three-hour drive south of Seattle, Washington.

www.travelportland.com

One of the 500-plus food carts (food trucks) that are scattered all over Portland

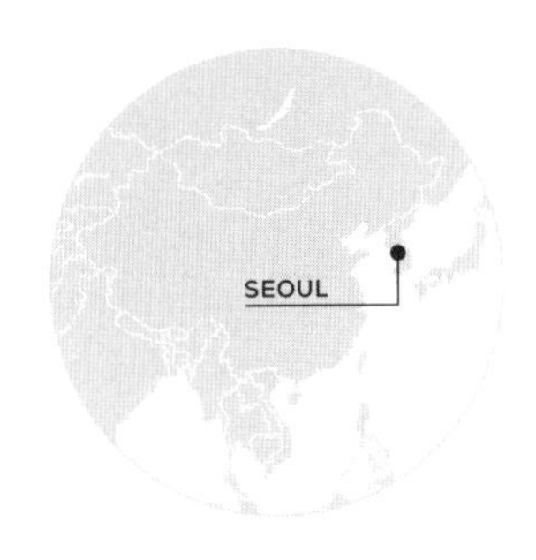

Seoul is a spirited dynamo of food, religion and culture. These things it shares with Tokyo, but while the Japanese capital barely looks up from the bustle, Seoul takes time to linger.

the alternative to Tokyo, Japan

SEOUL

South Korea

The present is a mirror to the past in Seoul, South Korea's irrepressible capital city, where ancient temples are reflected in the glass walls of space-age skyscrapers. In its juxtaposition of a vibrant future and an honoured past, Seoul bears much similarity to Japan's capital, Tokyo, but this is a quieter kind of place, with fewer tourists and, crucially, somewhere your hard-earned money goes a lot further.

Like Tokyo, Seoul is a haven for food lovers. While the Japanese delights of sushi, sashimi and yakitori have long appeared on restaurant menus the world over, a food journey through Korea's capital is more likely to be one of novelty and discovery. Dive into the chaotic, steam-wreathed alleyways of Gwanjang Market in pursuit of *mayak kimbap*, seaweed rolls stuffed with pickled radish and carrots, and you'll immediately understand why these moreish morsels are nicknamed "narcotic rice rolls". Kimchi is another of Korea's most addictive food offerings. Take a class at the Seoul Kimchi Academy and you'll leave wondering how you ever survived without this fermented cabbage marvel – luckily you'll now have the skills to create it at home.

Body well-nourished, turn your attentions to the soul. Age-old spirituality infuses the streets of Seoul. It's impossible not to be moved by Gyeongbokgung Temple, for example, which has been sacked, burned and destroyed, but risen renewed from the ashes every time. Korean religion continues to be influenced

The ornate Geunjeongjeon (Throne Hall) of the Gyeongbokgung Temple, in the historic heart of the city

Exploring the variety of food on offer at the bustling Kwangjang Market, a traditional street food market in Seoul

by shamanism. Like Shinto in Japan, this is an animistic religion, where ancestors and nature gods are venerated at atmospheric shrines. One thing you'll never find in Tokyo, though, is the delicate exorcism dance of a *mudang* (female shaman), a regular sight at Seoul's 14th-century hilltop shrine, Inwangsan – the perfect place to watch the sun set on an exciting day in the city.

Still going to Tokyo?
Tokyo's sardine-like crowds are legendary, and they're generally a year-round fixture. For a little respite, the quieter months of autumn or winter are good times to visit.

Getting There *Incheon International Airport connects Seoul to cities around the world.*

www.visitseoul.net

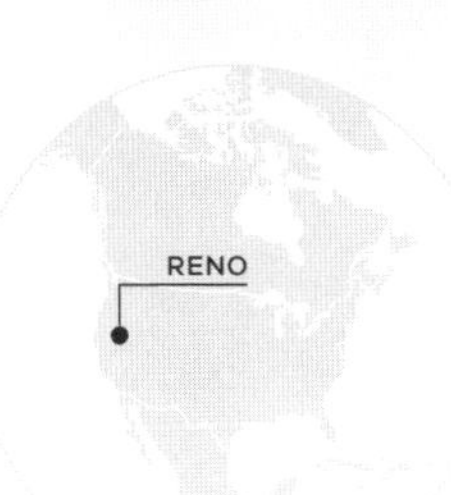

Sin City has the desert, but Reno has the Sierra Nevada, a mountain playground just outside the city. Of course, there's high-stakes gambling and late-night revelry on offer here, too.

the alternative to *Las Vegas, USA*

RENO

USA

Like Las Vegas, Reno has casinos – nearly two dozen squeezed into this tiny city – along with a buzzing bar scene. What gives Reno the edge on Vegas, however, isn't its gambling scene but its opportunities for outdoor adventures. Nestled on the edge of the Sierra Nevada, Reno offers hiking and mountain-biking in the mountains, skiing at spectacular winter resorts and even white-water rafting along the Truckee River. And while the summer brings hot days, they don't compare to the scorching temperatures of Vegas. With four seasons of activities, and sparkling Lake Tahoe just an hour away, there's really no bad time to visit Nevada's lively little city.

Still want to visit Las Vegas?
You'll find good room deals and fewer crowds if you visit mid-week or during the summer and winter low seasons.

Getting There *Fly into Ren-Tahoe International Airport, just 3km (2 miles) from the city. For activities outside the city, hire a car.*

www.visitrenotahoe.com

The capital of the North of England has the museums and music to rival the capital in the south – and it doesn't take itself so seriously. Whether you're after frivolous fun, cool architecture or cutting-edge art, you'll find it here.

***the alternative to** London, UK*

MANCHESTER

UK

London is crammed with iconic landmarks and world-class museums, and has a cosmopolitan dynamism that's matched by few other countries' capitals. But it's also a fact that most tourists to England wind up in London at some point. When you add that to the capital's huge population and the vast number of commuters piling into the city each day, it's no wonder the streets are chock-full and the queues long.

It's time for some levelling up. Look north to Manchester, England's second city and one of the most exciting places in the UK. The birthplace of the Industrial Revolution, Manchester grew rapidly in the mid-18th century, principally off the back of its flourishing cotton industry. The architectural legacy of this is a city that's fascinating to stroll around, with a network of canals and mills, and grids of streets whose appearance has been shaped by textile warehouses and Victorian markets.

At the Science and Industry Museum you can see the world's oldest surviving passenger railway station. And inside some of those red-brick mills you'll now find galleries with forward-thinking works by famous artists or, in the case of Islington Mill, up-and-coming local talent. The city's cultural scene is as vibrant as you'd expect from the self-styled Capital of the North. London may have the Tate Modern and the Barbican, but you'll always find a thought-provoking exhibition at the Manchester Art Gallery or The Whitworth.

Browsing Pre-Raphaelite paintings in the Manchester Art Gallery, a major museum of European art

Red-brick Victorian and modern architecture, standing side by side in the centre of Manchester

It's not all about museums, of course. Since the 1990s "Madchester" years that put the city's night clubs on the map, Manchester is *the* place for club nights. Night outs here are long and fun, helped in part by Mancunians' friendliness, the pride of the city. And with the Peak District National Park at its doorstep and the Lake District only an hour's drive away, Manchester is very close to some of the UK's most stunning natural landscapes.

So, look north to Manchester. The BBC did when it shifted half of its operation from London to the Salford Quays area of the city. And if it's good enough for the Beeb, then it's good enough for us, too.

Still going to London?

Avoid the school holidays. The city is a cultural colossus so avoid the museums of South Kensington and check out instead Sir John Soane's Museum, Dennis Severs' House and the Leighton House Museum.

Getting There *Manchester Airport is located 13 km (8 miles) from the city centre. Trains from the airport take 20 minutes to reach the city.*

www.visitmanchester.com

MORE LIKE THIS

LEEDS
UK
Vibrant Leeds is the Foodie Capital of the North of England. Thanks to the Henry Moore Institute and the Leeds Art Gallery, it's also the heart of the Yorkshire Sculpture Triangle. Shoppers will love the Victorian shopping arcades.

LIVERPOOL
UK
This major northern port city has one of the finest collections of museums and galleries outside London. Its rich musical heritage (not just The Beatles) has also made it a UNESCO City of Music. It has an impressive number of parks, too.

NEWCASTLE
UK
A night out on the Toon is legendary. But Newcastle also has a 12th-century castle, contemporary art galleries like the BALTIC Centre, and the revitalized Quayside, a nightlife hotspot with trendy bars and restaurants.

Kyoto's temples, palaces and shrines may be more highly polished and better known, but those in South Korea's Gyeongju are just as enthralling, if not more so. Evoking bygone days as the capital of a noble culture, Gyeongju is a true living museum.

the alternative to *Kyoto, Japan*

GYEONGJU

South Korea

Koreans call it "the museum without walls", and strolling through Gyeongju's storied streets, you'll soon understand why. History echoes around every corner: here a Buddhist temple wreathed in incense smoke, there an enigmatic burial mound, relics of the city's former life as the capital of the Silla Kingdom. In its wealth of temples and shrines, and the gorgeous, centuries-old architecture, Gyeongju resembles Kyoto, a former imperial capital of Japan. But Gyeongju has a greater variety of architecture, with different neighbourhoods offering a window into contrasting eras of the past.

Enter Yangdong Village, for example, and you'll be forgiven for thinking you've stepped into the 15th century. The low-slung houses are topped with overhanging thatch resembling a mushroom's cap, villagers ferment kimchi in giant urns, and the vistas of blossoming lotus ponds and glittering green rice-terrace shelves look much the same as they did in the days of the Joseon dynasty.

As in Kyoto, the spirit of Buddhism infuses Gyeongju's historic sights, and it's nowhere more strongly felt than at Seokguram, a marvellously atmospheric grotto that is part of the Bulguksa Temple complex, on Mount Tohamsan. Carvings of Buddhist saints and Hindu gods line the walls of this hilltop hermitage, presided over by a mighty statue of the Buddha. The fact it was possible to build this mountain masterpiece in the 8th century is a monument to the might and ingenuity of the Silla Kingdom. The view might be the best in South Korea – come at dawn to watch the sun rising like a honeyed orb from the East Sea.

All that is recent history, though, compared to the vast burial mounds of Tumuli Park, Gyeongju's most famous and

The restored ancient Bulguksa Temple, high up on the wooded slopes of Mount Tohamsan

Below The royal burial mounds in Tumuli Park Bottom Strolling in the forecourt of the Seokguram Grotto, part of the splendid Bulguska Buddhist temple complex

mysterious sight. Worn smooth and topped with turf by the passing of millennia, the mounds are not natural and were in fact grand tombs, built for the kings of Silla going back more than 2,000 years. For a glimpse of pre-Buddhist culture, cross the threshold into the Noseodong tombs. Here paintings of mythical animals like sky horses are relics of a time when shamanism was the dominant religion.

Blinking into the light, you'll be struck by the proximity of these prehistoric tombs to Gyeongju's bustling shopping district. They represent something typically Korean: ancient history, reverberating through the streets of a place very much alive.

Still going to Kyoto?

To explore Kyoto at a quiet, particularly atmospheric time, visit in winter, when the ancient temples are often picturesquely frosted with snow.

Getting There *Flights arrive at Incheon International Airport in Seoul, from where high-speed trains take just over two hours to reach Gyeongju.*

www.gyeongju.go.kr

With ancient artifacts on every corner and famous wonders aplenty, Rome is a busy bucket-list favourite. When it comes to easy-to-find culinary delights, however, Bologna wins hands down – and history buffs won't be disappointed.

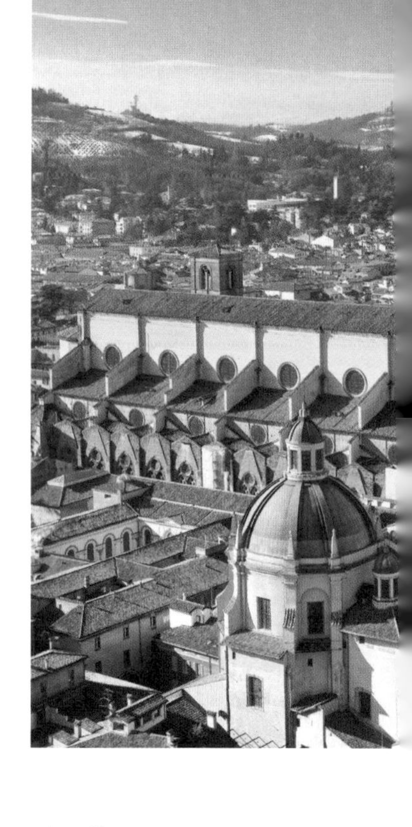

the alternative to Rome, Italy

BOLOGNA

Italy

Platters overflowing with freshly sliced mortadella and prosciutto, baskets of melt-in-the-mouth bakery bites, steaming bowls of pasta, washed down with refreshing local wines. With such glorious flavours on the menu, Bologna's title as Italy's foodie capital is nothing if not well deserved. While the culinary offerings are top-notch all over Italy (it's amazing what they do with offal in Rome), the seductively satisfying traditional dishes of Bologna really are a cut above the rest.

One of the best places for genuine flavours is the Quadrilatero market. Dating to the Middle Ages, this knot of narrow streets swirls with colour and aroma. Pavement tables provide perfect perches for savouring market-fresh munchies alongside stalls piled high with shiny fruit and veg. And just about any restaurant in town will serve *tagliatelle al ragù* (ribbon pasta with a rich meaty sauce). Places serving such authentic dishes are harder to find in Rome among the many touristy restaurants.

Of course Rome has knock-out buildings everywhere you look, but you'll be surprised what architecture you'll see on a post-lunch walk around Bologna. And unlike the Eternal City, with its seven hills and many neighbourhoods, Bologna is flat and easy to get around. Next to the market is Piazza Maggiore, surrounded by majestic medieval palazzi and the stately San Petronio basilica. This vast church was originally intended to be even bigger than St Peter's in Rome. Pope Pius IV cut the project

Shopping for fresh produce in the Quadrilatero, Bologna's oldest market

Left *The historical centre of Bologna with its medieval buildings* ***Below*** *Walking along one of Bologna's porticoes, which are a UNESCO World Heritage Site*

short, though, in favour of constructing the Archiginnasio next door, the first building to house all the city's university faculties.

Another key feature of Bologna are its old porticoes (covered pavements), which make roaming a joy even during the wettest winter weekend or hottest summer afternoon. You can walk along the world's longest continuous portico to the hilltop church of San Luca. What better way to work up an appetite for tasty delights?

Still visiting Rome?

Head south to the Castelli Romani area, which is famous for its food and wines.

Getting There *The Marconi Express monorail from Bologna Airport takes about eight minutes to reach Bologna Centrale station in the city centre.* ***www.bolognawelcome.com***

MORE LIKE THIS

RAVENNA
Italy
This small but important city stepped into Rome's shoes as capital in AD 402. It's the stunning Byzantine mosaics, along with its beaches, that are the real draw today.

PALERMO
Italy
As capital of Sicily, the seaside city of Palermo overflows with sights. Some of the most striking and decorative date from the city's Arab-Norman period.

Paris isn't the only city in France where you'll find stately boulevards, grand squares and galleries galore. La belle vie *is just as sweet in the south, where Montpellier will win you over with its Languedoc charm.*

***the alternative to** Paris, France*

MONTPELLIER

France

With its elegant promenades and honey-hued Haussmannian buildings, Montpellier is said to be Paris's smaller southern counterpart. Even the triumphal arch is a straight 17th-century copy of the Parisian original. In this case, imitation might be the sincerest form of flattery, but architecture is where any Parisian pastiche ends.

Montpellier marches to its own beat – and you'll have plenty of fun here. Not in fancy restaurants or exclusive clubs mind, but on *terrasses* lit by twinkling fairy lights or over cheap pints of beer in studenty dive bars. The food is hearty and the welcome warm, fuelled by robust red wines from the Languedoc and the rich flavours of garlic, thyme and rosemary.

Banish visions of Parisian shades of greige. Colour is everywhere, in murals and street art. Even stepped medieval lanes are decorated with a lick of rainbow paint and homemade bunting. This love of aesthetics is reflected in Montpellier's contemporary arts hub MOCO, which put the city on the map for avant-garde shows. And the Musée Fabre overflows with classical masterpieces by artists linked to the region – as well as works by the likes of Rubens, Delacroix and Monet. Artists are rightfully revered, but unlike in somewhat haughty Parisian circles, southern art connoisseurs tend to take themselves less seriously.

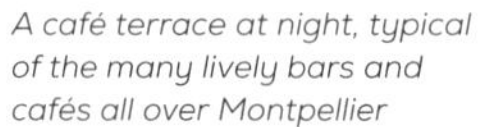

A café terrace at night, typical of the many lively bars and cafés all over Montpellier

Swathes of green are never far away. Montpellier's shady Jardin des Plantes is not only the oldest botanical garden in France but by far one of the most beautiful, and a more tranquil place to stroll than its younger Parisian sibling.

Of course, escaping to the Med from Montpellier couldn't be easier. The beaches of Carnon and La Grande-Motte are an easy 30-minute tram-ride away. Travel the same distance in Paris and you'd be stuck in traffic jams on the *périphérique*. On a summer afternoon, even Parisians would find it hard to argue they've got the better deal.

__Above__ The 1663 Porte du Peyrou triumphal arch, near the centre of the city __Right__ Beautiful Haussmann-style buildings, typical of Montpellier's architecture

Still going to Paris?

Visit in autumn rather than spring for crisp mornings crunching through fallen leaves in the Tuileries and afternoons at crowd-free museums. What you'll miss out in *pique-niques*, you'll make up for with sizzling raclette and Beaujolais nouveau.

Getting There *Montpellier Airport is 10 km (6 miles) outside the city; direct trains from Paris take three-and-a-half hours.*

www.montpellier-france.com

Swap the clamour of Delhi for the serenity of Udaipur. This lake-filled, picturesque city has no shortage of splendid palaces, temples and old twisty streets packed with bazaars.

Lake Pichola with the grand City Palace and other majestic architecture of the city of Udaipur

the alternative to *Delhi, India*

UDAIPUR

India

Leave the chaos and cacophony of Delhi behind, fascinating though India's capital is. Further south, Udaipur, the historic capital of the kingdom of Mewar in the large state of Rajasthan, is a far more tranquil city, with shimmering lakes and altogether cleaner air. Surrounded by lush hills and filled with resplendent maharajas' palaces, this 16th-century jewel is one of India's most romantic cities.

To say that Udaipur sits next to a lake is an understatement: it sits by two big lakes with a couple of smaller ones in between, and more scattered around it. The largest, Lake Pichola, gives the city its most dreamy allure, providing an idyllic backdrop to picture-postcard views. You can take a boat ride into the middle, and forget the city altogether (something not so easily done in Delhi).

Sitting right at Lake Pichola's edge, the grandiose City Palace, with its rows of arched windows and intricate turrets, contains a mesmerizing museum. Exploring this miniature kingdom of royal apartments, reception halls and courtyards, linked to each other by narrow passages and steep staircases, can take a whole day. Behind City Palace stretches Udaipur's old city, with its original walls, punctuated by huge ornate gates. The labyrinth of winding, hilly streets teems with bazaars, shops and old temples. After you've had your fill of these, Udaipur has some fun museums. The Vintage Car Museum has a top collection of antique royal cars, while the Crystal Gallery is full of exquisite, finely cut glass – we're talking not just tableware, but furniture too.

That's far from being the city's grandest display of opulence though. Old palaces that have been converted into super-deluxe hotels grace a couple of islands on Lake Pichola; anybody who stays here is guaranteed

The Badi Mahal (Garden Palace) with intricately carved archways, standing at the highest point of the City Palace

to feel like a maharaja. For those whose budgets don't quite stretch as far, a number of very fine but less expensive old *haveli* (stone mansion) hotels are scattered around the city. You can absorb stories about old buildings like these and the Kingdom of Mewar at the sound-and-light show, held in the grounds of the City Palace every evening. It's a hugely atmospheric and romantic spectacle.

Still visiting Delhi?

Take a guided walking tour of Old Delhi to learn about the oldest sights, many of which are actually older than the city itself.

__Getting There__ A taxi from Udaipur's Maharan Pratap Airport takes about 30 minutes to reach the city.

www.udaipurtourism.com

Often referred to as Vienna in miniature, Graz is Austria's second-largest city and a lively student hub. Come to enjoy a vibrant arts scene, stay for the exceptional regional cuisine.

***the alternative to** Vienna, Austria*

GRAZ

Austria

Located around 200 km (124 miles) south of Austria's capital, laid-back Graz feels a world away from traditional Vienna. Yes, there's still old-world grandeur to be found here (Graz's old town is a World Heritage Site) but in between princely palaces and Renaissance courtyards sit innovative creations, like the striking, conical Science Museum and the bulbous, alien-esque Modern Art Museum, a conversation starter if ever there was one.

And if you want something delicious to eat while you debate the merits of this UNESCO City of Design, you're in luck. Graz is also the country's culinary capital. Not to be missed is the local pumpkin seed oil, used in a plethora of dishes. Try it at some of the city's cool restaurants or grab a bottle from a farmers' market to take home. Oh Vienna? Oh Graz!

Still going to Vienna?

For a special view, head to the banks of the Old Danube to watch the sunset over the city skyline.

__Getting There__ Buses from Graz Airport take 30 minutes to reach the town. Trains from Vienna to Graz take two-and-a-half hours.

www.graz.at

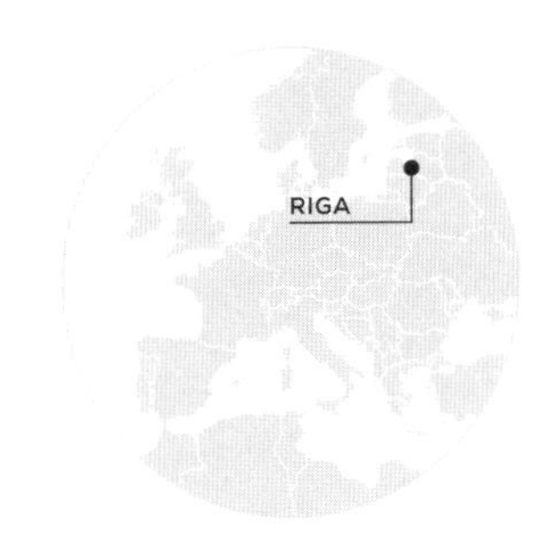

Like Prague, Latvia's capital offers a blend of medieval streets and 20th-century architecture, alongside superb museums and nightlife. Unlike Prague, it also offers spectacular sandy beaches within easy reach of the city's historic centre.

***the alternative to** Prague, Czech Republic*

RIGA

Latvia

With its UNESCO-listed Old Town chock-full of medieval, Gothic and Baroque buildings, Riga is often dubbed "the new Prague". From a high vantage point, there are clear parallels, with the city's stunning sweep of church spires, orange roofs and narrow streets set alongside a wide expanse of river. But a short stroll around the streets of the Latvian capital will quickly reveal a different picture. What's remarkable about this city isn't so much its centuries-old buildings, but its turn-of-the-20th-century ones: Riga is a hub of Art Nouveau architecture.

Prague may have given the world Art Nouveau icon Alphonse Mucha, but Riga is the city that really took the movement to its heart. As it underwent large-scale expansion in the late 1800s and early 1900s, vast swathes of the city's New Town were designed in this in-vogue style. There are eye-catching examples around almost every corner – don't miss the distinctive peach-coloured German Embassy, the ornate blue-and-white Mikhail Eisenstein mansion and the elegantly decorated Riga Synagogue. There's even an Art Nouveau Centre here, housed within what was once the private apartment of Konstantīns Pēkšēns, a prominent architect of the movement. Step inside this museum to see its spectacular spiral staircase (one of the capital's most photographed sights).

As a rule of thumb, "step inside" is good advice, as what lies within Riga's buildings is often just as fascinating as what is outside. This city has a host of world-class museums, including the Latvian War Museum (which tells the story of the country's conflicts over the years, particularly the 1918–20 War of Liberation) and the Open-Air Ethnographic Museum (which features homesteads from around the country).

While its museums are world-class, Riga's nightlife scene is also something to write home about. Long the most lively of the Baltic capitals, Riga is home to an array of pubs, cocktails joints, jazz bars and techno clubs, all of which keep the streets buzzing well into the early hours.

Clockwise from left
Latvia's picturesque cityscape, bordering the Daugava River; Mikhail Eisenstein's Art Nouveau mansion; performers dressed in traditional attire

This eastern European gem has an extra ace up its sleeve: the ocean. Miles of sandy shoreline skirt the bracing Baltic Sea on the edge of the city, while just a little further west lies Jūrmala, a popular resort town with its long stretch of white sand. Riga is defined by its proximity to the sea, something its Czech counterpart cannot compete with. So, while these two pretty cities have a lot in common, Riga is not the new Prague. It's so much more.

Still going to Prague?

After ticking off the big-ticket sights in the Old Town, head to one of the city's lesser-known, up-and-coming neighbourhoods such as verdant Vinohrady, industrial Karlín or eclectic Holešovice.

Getting There *Riga International Airport is the largest in the Baltic states. Riga's centre is very walk-able, while buses and trams connect the outer districts.*

www.liveriga.com

MORE LIKE THIS

KOŠICE
Slovakia
This small city in eastern Slovakia is full of stunning old buildings, including a magnificent medieval cathedral and a Neo-Gothic palace.

OLOMOUC
Czech Republic
It's not all about Prague. This eastern Czech city also boasts a gorgeous Baroque square with astronomical clock, as well a Holy Trinity Column – and barely a tourist in sight.

Like Mexico City, Zacatecas is full of historic sights and cultural gems. But it's also quieter and more compact than the sprawling, hectic capital, which makes exploring its heritage an easier, relaxing experience.

***the alternative to** Mexico City, Mexico*

ZACATECAS

Mexico

Wedged between arid hills some 2,440 m (8,010 ft) above sea level, Zacatecas is a jewel of a city. Mexico City might have myriad attractions, but it can be a challenge to enjoy them when you consider how busy, noisy, congested and sometimes chaotic the capital is. By contrast, Zacatecas' historic centre, which is lined with elegant streets and dotted with plazas, is comparatively calm and relatively easy to stroll around – an idyllic combination for a city break, if you ask us.

Founded in 1546 following the discovery of vast silver deposits, Zacatecas swiftly became a highly prosperous and politically significant place. The wealth flowed throughout the 16th, 17th, 18th and early 19th centuries, resulting in the construction of countless attractive churches, palaces, mansions and municipal buildings. The capital of Zacatecas state was the very first city in Mexico to undergo a rigorous preservation programme, and it shows. A UNESCO World Heritage Site since 1993, the downtown area is a faithfully restored Spanish colonial-era city like no other in the country. If that wasn't enough to shout about, it's home to a richly decorated 18th-century cathedral that is perhaps the finest example of Mexican Churrigueresque architecture – a flamboyant offshoot of the more familiar Baroque style – in the world. Even the most jaded traveller will be impressed by this soaring spiritual monument and its magnificent pink-stone façade.

These architectural marvels aren't just static buildings to admire. A number of former convents, palaces and mansions have been converted into first-rate museums and galleries showcasing local, national and international artists' work.

Woman in folk costume in front of the ornate entrance to the cathedral

Above *Brightly painted houses in the old city centre* ***Right*** *The elegant rooftops and domes of Zacatecas, spread out around the cathedral*

Take the famous royal mint, the Casa Real de la Moneda, which is now a museum. Or the Museo Rafael Coronel, which exhibits pottery from all over the world, as well as the largest collection of works by painter Joan Miró outside of Spain.

Zacatecas's vivid cultural scene doesn't stop at art. Quiet it may be, but when its numerous festivals and events come around every year, the city turns up the energy. The most vibrant is the Festival Cultural Zacatecas, a varied programme of plays, concerts, exhibitions and performances that takes place around Semana Santa (Holy Week) in March or April. There's more, too: a major folk festival, an impressive *charrería* (a traditional display of horsemanship), hot-air balloon festivals, street theatre extravaganzas, re-enactments of ancient battles and centuries-old religious processions that have long since died out elsewhere. Better still, most of these events are delightfully free.

Holding onto its legacies certainly defines this city. Just north of the centre is the legendary Mina El Edén, the source of the silver that provided Zacatecas – and, until Mexico won its independence, the country's Spanish colonial rulers – with immense wealth. The cost of these riches was borne by the region's Indigenous peoples, who were dispossessed of their lands, subjugated and left to toil in backbreaking conditions. The mine is now a museum, taking you on a railroad journey into the bowels of Cerro de Grillo, before exploring the tunnels, shafts and galleries that once echoed with the sound of hundreds of workers. Near the entrance, a cable car climbs up the Cerro de la Bufa, a hill that was the scene of a bloody battle in 1914 that ended with a victory for revolutionary

leader "Pancho" Villa over federal troops. At the top, the Museo Toma de Zacatecas is dedicated to the revolution.

And wait, there's more. Zacatecas' central location makes it an ideal base for trips to the archaeological ruins of La Quemada and quaint villages in the surrounding countryside. And with the Pacific and Gulf of Mexico coastlines close by, the country's gorgeous beaches are only ever a day trip away.

Still going to Mexico City?
Although some travellers find the hustle and bustle overwhelming, the capital's superb architecture, Aztec ruins and unrivalled cultural scene ensure it remains on many itineraries. Head into the historic centre early before the crowds build up.

Getting There *The city has an airport, Zacatecas International Airport, and a bus station serving destinations across Mexico.*

www.zacatecastravel.com

Top *Local street life in Zacatecas*
Above *The 17th-century Virgen del Patrocinio church*

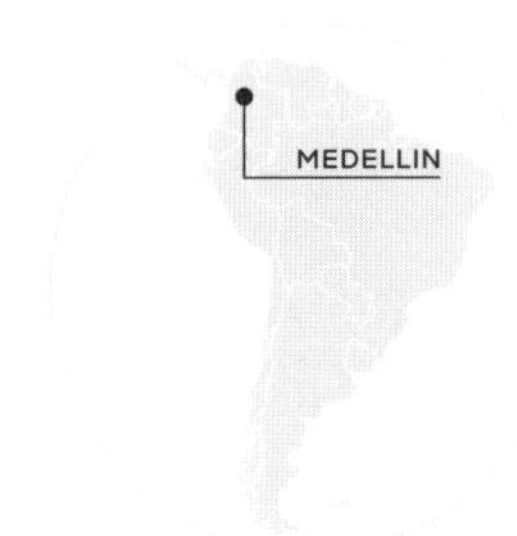

For a fun-filled South American city break, leave Rio behind and head to the fertile hills of Medellín. This is one of Colombia's most exciting urban centres, with great food and a buzzing nightlife scene.

the alternative to
Rio de Janeiro, Brazil

MEDELLIN

Colombia

High-spirited Rio may be the first South American city on your bucket list, but Brazil's vibrant centre has an even cooler competitor, Medellín. Cradled in the lush Aburrá valley, Colombia's creative second city is always ready to let its hair down.

There are two sides to Medellín. In the chic neighbourhood of El Poblado, lively restaurants serve up high-end Colombian fare, while a riot of bars see patrons *rumbiar* to European DJs and *cumbia* beats. This is party-loving Rio at a more intimate level, with young revellers keeping the clubs buzzing well into the early hours.

But board Medellín's modern metro – an emblem of the city's rebirth from the shadow of drug violence – and you're transported downtown, into the heart of local Medellín. Here, traditional, no-frills restaurants throng with locals slurping up hearty *cazuelas* (stews) and *paisa* soups. At Plaza Botero, the humorously rotund sculptures of Medellín artist Fernando Botero stand alongside coffee hawkers and markets. The buzz is palpable at every turn: this is a city emphatically proud of what it's become – and one you should upgrade to the very top of your bucket list.

Plaza Botero, filled with sculptures and greenery, next to the Iglesia de la Candelaria

Still going to Rio?

If visiting during Carnival, be sure to book tickets for the Sambadrome parades in advance to avoid missing out on one of the highlights of the festivities. Five days of street parties not your thing? Consider Rio during April or October for lighter crowds and cheaper prices.

Getting There *Fly into José María Córdova International Airport; from here it's a 25-minute taxi or an hour-long bus ride into the centre of Medellín.*

www.colombia.travel

Snuggled in the foothills of Haute-Savoie, with canals dramatically framed by the Alps, the French town of Annecy is one of the most romantic spots in Europe and a beautiful alternative to over-touristed Venice.

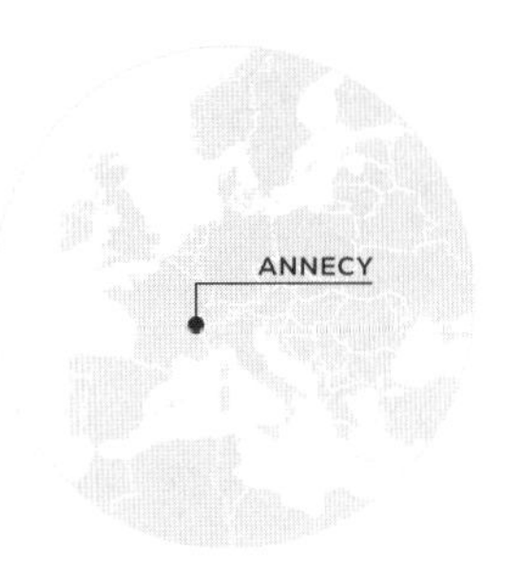

***the alternative to** Venice, Italy*

ANNECY

France

Dusted by snow in winter or dressed with colourful flower boxes in spring, Annecy enchants. Just like in Venice, winding canals snake between ancient buildings that lean towards the water's edge – only here they're not fed by a lagoon but by crystal-clear Lake Annecy. Gondolas don't float beneath Annecy's bridges either, just the occasional swan, which leaves visitors free to enjoy the scene uninterrupted by floating serenades.

Annecy is one of the rare places that really lives up to its fairy-tale reputation. Ancient, echoey churches and quaint, pastel-coloured buildings, complete with worn window shutters, line the cobbled streets of the old town. Just beside the town, the stunning Lake Annecy – a favourite swimming and paddle-boarding spot for locals – is framed by forested mountains. There's even a medieval castle, Château d'Annecy, overlooking the lake. Combine all of this with Annecy's location (just an hour's drive from the Swiss city of Geneva) and it's easy to see why the picture-perfect town was voted one of the best places to live in France. As for its tourist numbers? Its small size still keeps its popularity under wraps...for now.

A picturesque canal, lined with colourful buildings, against a backdrop of the northern French Alps

Still going to Venice?

If you go, go slow. Travel out of season, taking time to explore the mainland as well as the islands, and seek out local guides and businesses.

***Getting There** Geneva is the closest airport; direct trains from Paris take around three hours and 45 minutes.*

www. lac-annecy.com

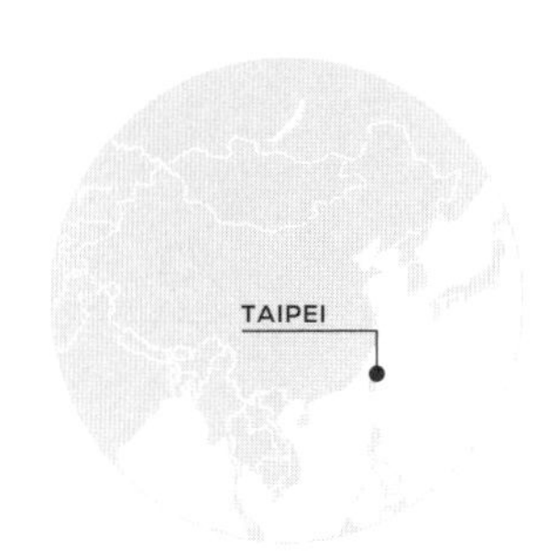

Hong Kong may have a perfect mix of tradition and ultra-modern, but you're more likely to be elbowing crowds out of the way to see it all, or rushing through on a layover. Taipei offers just as divine a blend at a more gentle pace.

the alternative to Hong Kong, China

TAIPEI

Taiwan

A metropolis where nature is never more than a few minutes away, Taipei offers as much to outdoor enthusiasts as it does to history buffs and foodies. While Hong Kong's dense skyscrapers huddle together like penguins, Taipei's relatively squat, heritage-rich neighbourhoods sprawl across the Taipei Basin. These cosy thoroughfares are overlooked by the massif of Yangmingshan, a national park just north of the city centre that features lush hiking trails, hot springs and a giant sleeping volcano.

Remnants of history lie scattered across Taipei, and its old neighbourhoods are wonderfully preserved. Amble over to Qidong Street and you'll discover a maze of charming lanes home to single-story Japanese colonial-era homes. Ancient temples, too, have withstood the test of time. Lungshan Temple, constructed in 1738, is the spiritual heart of the city and acts as a window into Old Taipei – with traditional shops and food stands hugging its perimeter. As you wander around this historic corner, you'll smell the heady aroma of barbecued squid and freshly made pepper buns – a bite of which will have you flabbergasted that Taipei's culinary delights are oft-overlooked on the global stage. Night markets across Taipei are just as rich with old-school dishes, from taro ice to seafood porridge.

In Hong Kong, old and new are jostling for attention, its glitzy shopping malls towering over smoky temples. Taipei's contrasts, meanwhile, appear to have more breathing room. Take Dihua Street,

The splendidly ornate Lungshan Temple in Wanhua District

Admiring the epic cityscape, sprawling out around the Taipei 101 skyscraper, from the verdant Elephant Mountain

lined with red-brick shophouses, which marks the place where Taipei first struck gold as a trade port along the Danshui River. A few stops east on the modern metro is Xinyi, a haven for glittering shops and luxury malls. Just a short walk south from this flashy district, the atmosphere changes again. The beloved hiking spot of Elephant Mountain offers views of Taipei 101 (a landmark skyscraper) all the way to the seaside suburb of Danshui and the sea beyond. Everything blends seamlessly.

Still want to see Hong Kong?

Avoid the urban crush in the city centre and venture to the lesser-known enclaves of Lamma Island. These hipster hubs are rife with hiking trails that promise coastal views and WWII-era Kamikaze Caves.

Getting There *Taipei can be reached by bus or the MRT's Airport Line from Taoyuan International Airport, Taiwan's main aerial hub.*

www.travel.taipei

Top *One of the many food vendors lining popular Danshui Old Street and Waterfront* ***Above*** *The strikingly modern Lover's Bridge, stretching over the Tamsui River in Tamsui Fisherman's Wharf*

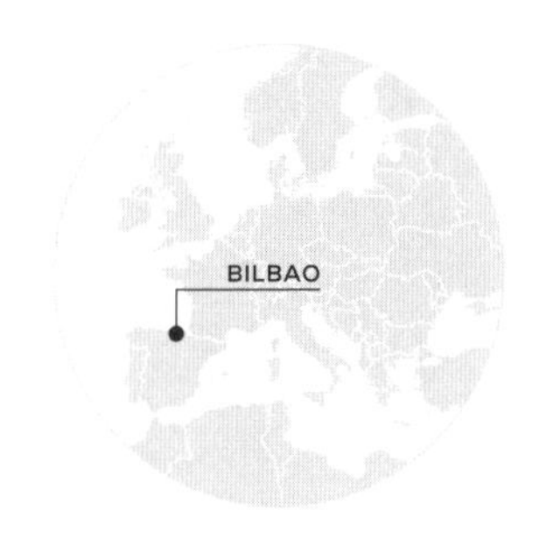

Madrid is Spain's age-old capital of culture, but Bilbao has brushed off its reputation as a gritty industrial port and is ready to take on the big guns. Head to this up-and-coming city for a weekend of alternative art and irresistible food.

Bibao's iconic Guggenheim Museum, guarded by Louise Bourgeois' Maman sculpture

the alternative to Madrid, Spain

BILBAO

Spain

Madrid is undoubtedly a powerhouse of cultural offerings, but Bilbao, the capital's smaller (and arguably cooler) cousin has just as much to offer art-lovers and foodies. Here, modern art and regional cuisine are championed, and Bilbao's all the more special for it.

Frank Gehry put Spain's northern port on the map in 1997 when he let his imagination run riot on the design of the spectacular Guggenheim Museum. Whereas Madrid's art museums are famous for their imperious collections – featuring Francisco de Goya, Diego Velázquez and Salvador Dalí – the Guggenheim draws visitors as much for the building itself as for the avant-garde works within. The structure swirls alongside the River Nervión and is clad in striking titanium panels. Inside, the focus is on contemporary and modern art, with changing exhibitions and a number of celebrated permanent works, like Jeff Koons' *Puppy* and Louise Bourgeois' spider-like *Maman*. Art in Bilbao isn't confined to its museums though. The city has become an urban gallery for street artists from around the world, and a wander through the neighbourhood of Bilbao La Vieja and out to Plaza Kirikiño will take you past renowned murals by Erb Mon, Sixe Paredes and DK Muralismo.

Food has also played a central role to Bilbao's revival. This city is a hotspot for foodies, with five times as many Michelin-starred restaurants per capita

Locals relaxing in Plaza Nueva, the scenic square in the centre of Bilbao famous for its numerous pintxos bars

as there are in Madrid. For those who don't dig fine dining, there are the simpler *pintxos*, the delicious Basque equivalent of tapas, piled high on plates atop bar counters all around the city. Why not combine Bilbao's two great loves on a *pintxos* bar crawl, eating your way around the city, passing colourful murals and street art as you go?

Still going to Madrid?
Madrid is a furnace in the summer, so visit in the spring or autumn. You can save money on entry to the big art museums by buying a Paseo del Arte combined ticket.

Getting There *The city centre is a 25-minute bus ride from Bilbao Airport. Alternatively, you can hop in a taxi or hire a car.*

www.bilbaoturismo.net

PERTH

Like its east coast fellow, Perth promises golden beaches, stunning scenery and a burgeoning arts scene. We'd take Western Australia's sunny capital over busy Sydney any day.

the alternative to *Sydney, Australia*

PERTH

Australia

Isolated from its easterly urban counterparts – and the rest of the world for that matter – Perth features the perfect blend of beautiful landscapes, enticing cultural offerings and a famous unhurried charm. Here, you'll find brochure-blue waters and endless stretches of sandy beaches – the most famous of which, Cottesloe, can easily rival Sydney's Bondi Beach. A quick boat ride away is Rottnest Island, a haven for bike lovers and smiling quokkas, while inland, numerous state forests and prestigious wine regions provide great day trips from the city.

Perth's cultural scene is just as exciting: think buzzing bars and restaurants, live music venues and street art murals hidden down little lanes, plus the laugh-out-loud Fringe World Festival that adds to the unique west coast vibe that is just "so Perth".

Still going to Sydney?
To make the most of Sydney's coastal lifestyle, visit in early autumn – so you get the tail end of hot days plus off-peak tourist numbers.

Getting There *Buses connect Perth's international airport with the city centre within half an hour.*

www.visitperth.com

INDEX

A

B

C

D

E

F

Q

R

S

T

U

V

W

Z

ACKNOWLEDGMENTS

Dorling Kindersley would like to thank the following contributors for their words:

Rob Ainsley writes about cycle touring in books and magazines. He collects international End-to-End rides and blogs on e2e.bike. His favourite place to tour is the next one.

Eleanor Aldridge is an author and journalist specializing in food, travel and (often natural) wine. She's been based in Paris for four years, writing about French culture and the city's little-known highlights.

Flora Baker is a writer, blogger and author based in London. She runs the award-winning travel website *Flora The Explorer* and has written for *Coastal Living*, *The Telegraph*, and *National Geographic Traveller*.

Ros Belford has written numerous guidebooks to Italy, and also writes on Italy and Sicily for *The Telegraph* and *Condé Nast Traveller*. After living full time in Sicily for 12 years, she now divides her time between Cambridge, Cornwall, Siracusa and the Aeolian Islands.

Julia D'Orazio is a travel writer based in Perth, Australia. Besides eating her way through over 70 countries, she counts Eurovision and scuba diving as her two greatest loves.

Keith Drew is a former Managing Editor at Rough Guides. He has travelled to 70 countries and counting and writes about his adventures for *The Telegraph* and *BBC Travel* among others. He is the co-founder of family-travel website Lijoma.com, a selection of inspirational itineraries.

Steph Dyson is a bilingual travel writer and journalist specializing in sustainable travel with a penchant for adventure and a passion for all things South American.

Marco Ferrarese has lived in Penang since 2009, from where he covers Malaysia, the Indian Subcontinent, China and the larger Southeast Asian region for DK and Lonely Planet and for *BBC Travel*.

Robin Gauldie has travelled widely in Europe, Africa, Asia and the Americas. He is the author of numerous travel guidebooks for DK and other publishers, including *DK Eyewitness Top 10 Cyprus* and *DK Eyewitness Top 10 Crete*. He lives in Edinburgh.

Rob Goss is a Tokyo-based writer and author, covering travel and culture in Japan for publications such as *National Geographic*, *BBC Travel* and other media around the globe. He's also written seven books on Japan.

Joe Henley is a freelance writer, screenwriter, author and musician, based in Taipei, but frequently on the move for travel assignments, music and as many writing-related gigs as he can get.

Sophie Ibbotson read Oriental Studies and has spent 15 years working in Asia and the former Soviet Union. She specializes in emerging destinations, in particular in Central Asia and the Caucasus.

Gabrielle Innes is an Australian editor and travel writer based in Berlin. She has written for DK Eyewitness and Lonely Planet, among other publications.

Anita Isalska is a writer specializing in Australia, France and Eastern Europe. Born in Britain and based in California, Anita writes about road trips, food and offbeat travel. Read more: www.anitaisalska.com.

Daniel Jacobs is from London and has travelled widely in Europe, Asia, Africa and Latin America. He has contributed to a number of guidebooks on Morocco, Egypt, India, Kenya, Colombia and Brazil.

Kana Kavon is a trilingual writer of travel and educational content. She writes for children and adults, focusing on Central and South American history and culture.

Stephen Keeling has lived in New York City since 2006 and has worked on numerous titles for DK, including the guides to New York City, California, Florida and the USA.

Sarah Lane is a long-term resident of Bologna, particularly enamoured of Italy's amazing food and wine scene. She loves sharing the best local delights – both writing about them and running private tours.

Kate Mann is originally from London but has been living in Munich for many years. She writes about travel, food and culture in Germany and beyond – with words in *Condé Nast Traveller*, *BBC Travel*, Lonely Planet and more.

Shafik Meghji is an award-winning travel writer, journalist and author of *Crossed off the Map: Travels in Bolivia*. He contributes to *BBC Travel*, *Wanderlust* and Lonely Planet, among others.

Rachel Mills is a contributor to Rough Guides and DK Eyewitness guides to New Zealand, India, Canada, Ireland and Great Britain, and is an expert in sustainable tourism. Follow her @rachmillstravel.

Allan Mutuku-Kortbæk is an avid writer, photographer and marketeer who has lived in many places around the globe. For now Copenhagen is his home but the world is his oyster.

Jabulile Ngwenya is a South African travel coach and writer with a keen interest in African travel. Her passion is storytelling and documenting people, places and sunsets. Instagram: @travelstoriesafrica

Victor O'Sullivan writes travel guidebooks and freelances for *The Guardian*, *The Irish Times*, *Condé Nast*, *The LA Times*, *The Chicago Tribune*, *Travel & Leisure* and others. His Twitter handle is @VicBunratty.

Georgia Platman is a writer, nurse and teacher. Originally from London, she spent years travelling and living in South America. She now lives in Suffolk with husband John and daughter Joni.

Joseph Reaney is a travel writer based in Prague. He writes for DK Eyewitness, Lonely Planet, Fodor's and more, and also runs the travel content agency World Words *(www.world-words.com)*.

Daniel Robinson has covered France, Israel and Southeast Asia for a variety of guidebook publishers and periodicals, including *The New York Times*, for over three decades. His work has been translated into 10 languages.

Kristen Shoates is a writer and brand strategist based in Nashville, Tennessee. Always up for a new trip or adventure, she loves experiencing different cultures and telling the stories of people and places she visits.

Deborah Soden is a freelance writer based in Sydney, Australia. She has spent three decades crafting stories about travel, technology and business for book publishers, media outlets and corporate clients.

Regis St Louis has spent the past two decades exploring the lesser-known wonders of the world, from the Chihuahuan Desert in northern Mexico to the volcanic islands of Papua New Guinea. He's also contributed to over 100 travel guides, covering destinations on six continents. He currently lives in New Orleans.

Daniel Stables has authored or contributed to more than 30 travel books on destinations across Asia, Europe and the Americas. You can find more of his work on Twitter @DanStables, on Instagram @DanStabs, or on his website, www.danielstables.co.uk.

Lisa Voormeij resides in British Columbia and is a regular contributor to DK Eyewitness. Find her snorkelling with turtles in Hawaii, hiking the rainforests of the Pacific Northwest or seeking out unique local fare around the Mediterranean.

The publisher would like to thank the following for their kind permission to reproduce their photographs:

(Key: a-above; b-below/bottom; c-centre; f-far; l-left; r-right; t-top)

123RF.com: 4045qd 135, alzamu79 25cr, asiastock 112cr, bbsferrari 36t, bloodua 195cr, bwzenith 93cr, charles03 19tr, checubus 124tr, davidzfr 20-21, jakobradlgruber 96-7t, olli0815 52-3t, pjworldtour 19cr, shalamov 131cr, tupungato 113b

Alamy Stock Photo: 29tr, agefotostock / Javier Larrea 178-9, agefotostock / Kateryna Kolesnyk 96br, amnat 108, Tomas Anderson 93b, Al Argueta 30-31, Andre Babiak 75, D. Holden Bailey 157br, Antony Baxter 52b, Bildarchiv Monheim GmbH / Gerhard Hagen 173, Mel Birch 42, Eduardo Blanco 45tr, Blue Planet Archive CMA 156-7, Michael Brooks 184tc, Debu55y 194-5, Danita Delimont 164tl, Delphotos 69b, dpa picture alliance 171br, Giulio Ercolani 18b, Michele Falzone 209tc, Tiago Fernandez 16, FLPA / Alamy 156tl, Patrik Forsberg 214, Leo Francini 175, Gabbro 146-7tl, Ganesha 169cr, MARCOCCHI GIULIO / SIPA 55tr, Jinny Goodman 72, Diego Grandi 105, GRANT ROONEY PREMIUM 188, guichaoua 200bl, GUIZIOU Franck / hemis.fr 197b, Luis Gutierrez / NortePhoto.com 74-5c, Hufton+Crow-VIEW 171tr, Iconpix 165tl, Image Professionals GmbH / Ernst Wrba 47, imageBROKER / Christian GUY 65br, imageBROKER / Christian GUY 68-9tc, imageBROKER / Josef Beck 148, imageBROKER / Peter Giovannini 90-91, Ivoha 160br, Jeffrey Isaac Greenberg 13+ 169tr, Jeffrey Isaac Greenberg 4+ 168bl, Jon Arnold Images Ltd / Jane Sweeney 140-41tc, Jon G. Fuller / VWPics 165tr, Suzuki Kaku 162-163, Olga Khoroshunova 93tr, kud108 65tr, Sébastien Lecocq 136-7, Y. Levy 172, Christoph Lischetzki 86, Luz y Sombra 70-71, MAISANT Ludovic / hemis.fr 122, Don Mammoser 20, MAMO Alessio / hemis.fr 29br, markferguson2 149br, Stefano Politi Markovina 112-13tl, Terry Mathews 12cl, mauritius images GmbH / Jutta Ulmer 69t, mauritius images GmbH / Reinhard Dirscherl 154-5, Martin Meacnarowski 80tl, Carol Moir 46, Paula Montenegro 185tl, Ingo Oeland 134, Brian Overcast 75br, Peter Adams Photography 198, Kim Petersen 160bl, Marina Pissarova 34-5t, Prisma / Heeb Christian 164tr, Prisma by Dukas Presseagentur GmbH / Alamy 121tr, Sergi Reboredo 132-3tc, Sergi Reboredo 157tr, REUTERS / VINCENT WEST 45br, RIEGER Bertrand / hemis.fr 165cr, robertharding / Julian Elliott 106bl, robertharding / Matthew Williams-Ellis 137, David Robertson 149cr, Grant Rooney 115br, Ivan Sebborn 15tr, galit seligmann 61, Simon Dack News 43br, Smith Collection / Gado 124tl, Fredrik Stenström 23tr, Eileen Tan 26-7, The Hoberman Collection 170-71tc, Thye Gn 190-91, Tom Till 141cr, Steve Vidler 116, Darko Vrcan 98tl, Matthew Wakem 123b, Andrew Walmsley 51, Watchtheworld 175tr, WaterFrame_dpr 157c, WaterFrame_fba 141tr, WaterFrame_rok 157cr, Hilda Weges 179b, WENN Rights Ltd 48-9, Westend61 GmbH 131tr, Ray Wilson 146tl, Bruce Yuanyue Bi / Danita Delimont 106t, Bruce Yuanyue Bi / Danita Delimont 106br

AWL Images: 199, Danita Delimont Stock 87tr, Tom Mackie 118-19, Nigel Pavitt 130-31tc, Steve Vidler 194b

Depositphotos Inc: fedevphoto 99br

Dreamstime.com: Ammit 147cr, Animaflora 187cr, Askoldsb 205cr, Jon Bilous 184tl, Ryhor Bruyeu 205tr, Tsangming Chang 213cr, Paulo Costa 140, Songquan Deng 166b, Dudlajzov 197tr, Everst 73, Franz1212 187tr, Galinasavina 104, Lijuan Guo 139, Pablo Hidalgo 144-5, Paula Joyce 131br, Robert Jürges 99cr, Thomas Jurkowski 13br, Denis Kelly 56, Shahid Khan 195tr, Kosmos111 165b, Mirko Kuzmanovic 121br, Chayakorn Lotongkum 122-3t, Mariagroth 99tr, Markpittimages 210, Ben Mcrae 138-9, Giulio Mignani 76-77, MNStudio 81cr, Mrreporter 179, Onlyfabrizio 34bl, Yooran Park 196-7tl, Sean Pavone 166-7t, Anton Petrus 24-5, Photoprofi30 43cr, Jesus Eloy Ramos Lara 209cr, Luca Roggero 202br, Rosshelen 201cr, Saiko3p 25tr, Tyler Stipp 84, Tangducminh 113tr, Travelling-light 12-13c, Gorka Vega Barbero 81b, Wirestock 208-9tc

Getty Images: 1001slide 6b, AFP / CESAR MANSO 46-7c, AFP / FILIPPO MONTEFORTE 65cr, AFP / MIGUEL ROJO 40-41, AFP / NELSON ALMEIDA 59, AFP / PABLO PORCIUNCULA 41tr, Atlantide Phototravel 168-9, Gerrit Bril / EyeEm 130tl, Corbis News / Andrew Lichtenstein 169br, Corbis News / Paulo Fridman 174-175tc, Marco Cristofori 110-111, DigitalVision / Matteo Colombo 42-43, E+ / Global_Pics 150, E+ / THEGIFT777 185cr, Europa Press News 64, Alberto Gagliardi 55b, Richard I'Anson 41br, AFP / JACK GUEZ 60-61,

Julian Elliott Photography 198-9tc, Julian Elliott Photography 112-13tc, Amir Levy 60b, Loop Images 10-11, Reynold Mainse / Design Pics 128, Maremagnum 28tl, Martin Zwick / REDA&CO / Universal Images Group 82-3, Moment / By Bruce Bordelon 117, Moment / CR Shelare 22, Moment / Eric Bowers Photo 182-183, Moment / Michele D'Amico supersky77 81tl, Moment Unreleased / MB Photography 6tr, Ramiro Olaciregui 17, Eloi Omella 64-5, Anton Petrus 14-15tl, Andrea Pistolesi 114-15, Redferns / Caitlin Mogridge 51cr, Redferns / Joseph Okpako 50, Redferns / Joseph Okpako 50-51, Stockbyte / Christopher Kidd 78-9, Stone / Saha Entertainment 94-5t, Stone / tororo 94b, The Image Bank Unreleased / Richard I'Anson 6tl, Universal Images Group Editorial / Education Images 98-9tc, Universal Images Group Editorial / REDA&CO 80tr, WireImage / Josh Brasted 58, Zane Erasmus / EyeEm 138b

Getty Images / iStock: 1970s 120-21, 115tr, agustavop 143tl, Sergei Aleshin 108-9t, alvarobueno 19br, artiss 212-213, banjongseal324 23cr, BeNicoMa 146tr, bloodua 23b, Mubera Boskov 102-3, byheaven 202-3t, Roop Dey 119tr, Dhoxax 211, DoraDalton 137br, double_p 81tr, Reuber Duarte 150-51t, EJJohnsonPhotography 141br, ePhotocorp 149tr, ermess 36bl, font83 153, FotografieLink 186-7, fotoVoyager 77tr, fotoVoyager 192bl, Alberto Gagliardi 54-5c, IakovKalinin 189, Images_By_Kenny 213tr, Antonio Jacome 139cr, jjmm888 148-9tc, joningall 92tc, Ingus Kruklitis 28-9, Mirko Kuzmanovic 103br, Julia Kuznetsova 32, Leamus 13tr, Marcus Lindstrom 204-5t, Elijah Lovkoff 185br, MattGush 208tl, mazzzur 22t, Mlenny 33, Denis Moskvinov 43tr, no_limit_pictures 84-5, Onfokus 152, Ozbalci 18-19, Sean Pavone 185tr, PhotoMartin 13cr, piola666 92-3tc, RADZONIMO 97bl, Cheryl Ramalho 131tl, ROMAOSLO 56-7, RossHelen 200-201t, Marko Rupena 103tr, Nazar Rybak 143tr, SanderStock 92tl, SeppFriedhuber 119br, Shootdiem 86-7, Oleh Slobodeniuk 87br, StockByM 4-5, Subbotsky 143, tupungato 212, Julien Viry 192-3, YinYang 139tr, ZimbaX20 77br

Musée Jacquemart André: 176-7, Culturespaces - S. Lloyd 176bl

Rovos Rail: Jos Beltman 88-9t, DOOK 88b

Shutterstock.com: Andres Fernando Allain 62b, Paolo Bona 57, Inu 36br, Viacheslav Lopatin 160t, Angela Meier 62-3t, Mltz 15br, Glenn Perreira 124b, Ksenia Ragozina 132

Unsplash: Ryan Booth 195br, Jorge Fernández Salas 214-15t, Karen Z 191c, Abel Robles 206-7

All other images © Dorling Kindersley

Project Editors Rada Radojicic, Lucy Sara-Kelly
Editors Zoë Rutland, Elspeth Beidas
Senior Designers Tania da Silva Gomes, Ben Hinks
Designer Jordan Lambley
Proofreader Stephanie Smith
Indexer Hilary Bird
Senior Cartographic Editor Casper Morris
Jacket Designer Jordan Lambley
Picture Researcher Marta Bescos
Senior Production Editor Jason Little
Production Controller Kariss Ainsworth
Managing Editor Hollie Teague
Managing Art Editors Bess Daly, Sarah Snelling
Art Director Maxine Pedliham
Publishing Director Georgina Dee

First published in Great Britain in 2022 by
Dorling Kindersley Limited
DK, One Embassy Gardens, 8 Viaduct Gardens,
London, SW11 7BW

The authorised representative in the EEA is
Dorling Kindersley Verlag GmbH. Arnulfstr. 124,
80636 Munich, Germany

10 9 8 7 6 5 4 3 2 1
001-331751-Sep/2022

A CIP catalogue record for this book is available from the British Library.
ISBN: 978-0-2415-6883-5

Printed and bound in Malaysia

For the curious
www.dk.com

This book was made with Forest Stewardship Council™ certified paper – one small step in DK's commitment to a sustainable future. For more information go to www.dk.com/our-green-pledge

The rapid rate at which the world is changing is constantly keeping the DK Eyewitness team on our toes. While we've worked hard to ensure that *Go Here Instead* is accurate and up-to-date, we know that events get cancelled, sights close and trails become impassable. What's more, what was once off the tourist trail today could be everyone's favourite place tomorrow. So, if you notice we've got something wrong or want to reveal your own secret spot, we want to hear about it. Please get in touch at travelguides@dk.co.uk